Dust Explosions in the Pharmaceutical and Food Industries

This book provides a comprehensive and scientifically rigorous exploration of dust explosion hazards, with a specific focus on both pharmaceutical and food industries. It systematically examines the fundamental and applied aspects of dust explosion safety, integrating key concepts from physical chemistry, engineering, and risk management fields.

- Introduces the concept of dust explosions, defining their nature, causes, and most influential factors.
- Presents case studies and an in-depth review of significant accidents over the past 50 years in both the pharma and food industries, highlighting lessons learned and the ongoing challenges in preventing such accidents.
- Delves into the phenomenology of organic dust explosions, elucidating the physicochemical phenomena underlying the ignition and propagation of dust explosions.
- Addresses the experimental methodologies used to characterize dust explosion hazards.
- Explores risk mitigation strategies and protective measures tailored to the characteristics of the specific organic dusts and industrial equipment treated in the book.

By synthesizing the latest knowledge and addressing both theoretical and practical aspects, this book serves as a resource for a wide range of readers, including safety professionals, engineers, researchers, and students. It will not only equip readers with the tools to assess and mitigate explosion risks but also aim to inspire ongoing research and innovation in this critical area of industrial safety.

Sabrina Copelli, Associate Professor at the University of Insubria, is a researcher in the fields of Applied Physical Chemistry and Chemical Engineering. After studying Chemical Engineering, she specialized in Industrial Process Safety, earning her PhD in Industrial Chemistry and Chemical Engineering with honors at Politecnico di Milano. Professor Copelli's research spans a wide range of topics, with a primary focus on dust explosion modeling, chemical reaction kinetics studies, chemical reactor simulation, and thermo-chemical stability characterization of reactive mixtures.

Davide Ballinari began his working experience as a general worker in a factory for ten years. During that period, he decided to start his academic career as a working student, achieving his Bachelor's Degree in Safety Engineering in 2017. In the meantime, he moved to a Fire Reaction Testing Laboratory as an inspector and commercial technician, and he earned his Master's Degree in Environmental and Safety Engineering at the University of Insubria in 2023. Under the supervision of Prof Sabrina Copelli, he is a PhD student, specializing in Process Safety Engineering with particular focus on Dust Explosions and Calorimetry.

Dust Explosions in the Pharmaceutical and Food Industries

Sabrina Copelli and Davide Ballinari

Foreword by Chris Cloney

CRC Press
Taylor & Francis Group
Boca Raton London New York

CRC Press is an imprint of the
Taylor & Francis Group, an **informa** business

Designed cover image: © 2026 Shutterstock

First edition published 2027
by CRC Press
2385 NW Executive Center Drive, Suite 320, Boca Raton FL 33431

and by CRC Press
4 Park Square, Milton Park, Abingdon, Oxon, OX14 4RN

CRC Press is an imprint of Taylor & Francis Group, LLC

ISBN: 978-1-041-08425-9 (hbk)
ISBN: 978-1-041-08426-6 (pbk)
ISBN: 978-1-003-64532-0 (ebk)

DOI: 10.1201/9781003645320

Typeset in Times
by Deanta Global Publishing Services, Chennai, India

Contents

Foreword

Dust explosions remain one of the most persistent and devastating hazards faced by process industries. What makes this reality particularly troubling is not a lack of awareness but rather how long this hazard has been recognized and how frequently it has continued to manifest. Historical records show that between 1900 and 1956 alone, more than 1,000 dust explosions were reported in the United States. Subsequent decades have added hundreds more documented incidents across North America, Europe, and Asia, resulting in well over a thousand fatalities and several thousand serious injuries worldwide. These figures do not account for chronic underreporting, limited transparency in less developed regions, or the especially severe losses associated with coal mining operations globally that add thousands more fatalities when included. Despite advances in science, standards, and engineering controls, dust explosions remain a recurring feature of industrial accident statistics rather than a relic of the past.

Against this backdrop, the need for technically rigorous, industry-relevant, and forward-looking reference works has never been greater. This book addresses that need directly. Rather than treating combustible dust hazards as a narrow or isolated problem, it recognizes their complexity, variability, and deep integration into modern industrial systems.

This text focuses on two industries that continue to account for a significant portion of dust explosion incidents worldwide: food production and pharmaceutical manufacturing. The food sector remains one of the most heavily affected industries, owing to its scale, diversity of raw materials, and the widespread presence of finely divided organic dusts across milling, drying, conveying, and storage operations. Pharmaceutical manufacturing presents a different, and in many respects more challenging, risk profile. Facilities often handle highly energetic powders with very low minimum ignition energies, operate with hybrid mixtures involving solvents or vapors, and maintain ultra-clean environments that can obscure or delay hazard recognition. Addressing these realities requires more than generic guidance; it demands a nuanced understanding of both equipment-level hazards and system-level safety methodologies.

Another area covered in detail in this work is quantitative risk analysis. The text moves beyond purely qualitative assessments and provides a structured discussion of modern quantitative techniques as applied to combustible dust hazards. By explicitly linking risk assessment methodologies to the fundamental elements of dust explosion phenomena and by illustrating their application through realistic industrial examples, the book offers readers a practical framework for applying these methodologies.

Lastly, this textbook looks ahead to unresolved technical challenges that will shape future research and practice, from nanopowders and hybrid mixtures to scale-up effects and explosion venting optimization. By openly addressing uncertainty and complexity, and by examining the emerging role of tools such as artificial intelligence, machine learning, and the Internet of Things in supporting risk assessment, the authors encourage a mindset of continuous improvement as we move forward with these technological advances.

This book will be of particular value to engineers, safety professionals, and technical leaders working in food and pharmaceutical operations, as well as to consultants and ATEX or process safety specialists supporting multiple facilities across these sectors. Its depth of technical treatment, combined with its attention to real-world application, makes it a reference that will inform both day-to-day decisions and long-term safety strategies. In that sense, it represents a meaningful contribution to the ongoing effort to understand, manage, and ultimately prevent dust explosion hazards in modern industry.

Chris Cloney, PhD, PEng
DustEx Research Ltd
DustSafetyScience.com

1 Introduction

Combustible dust explosions remain among the most severe and underestimated hazards in process industries, particularly in sectors where fine organic powders are routinely handled, such as the pharmaceutical and food industries. Despite significant advances in process technology, automation, and regulatory frameworks, dust explosions continue to cause fatalities, serious injuries, environmental damage, and substantial economic losses worldwide. This book is devoted to providing a comprehensive, technically grounded, and application-oriented treatment of dust explosion phenomena, with particular emphasis on industrial reality, experimental characterization, risk management, and future developments.

The starting point of this work is the recognition that dust explosions are not rare or exceptional events, but recurring industrial accidents whose root causes often lie in insufficient hazard awareness, inadequate material characterization, and the misapplication (or absence) of appropriate prevention and protection measures. Chapter 2 addresses this issue by examining notable dust explosion accidents occurring in the pharmaceutical and food industries, drawing on documented case histories to illustrate the catastrophic consequences of combustible dust hazards. Through detailed analysis of events such as the "West Pharmaceutical Services" explosion and the "Imperial Sugar" refinery disaster, the chapter highlights common failure modes, including dust accumulation, a lack of hazard recognition, poor housekeeping, inadequate equipment design, and weaknesses in safety culture. Rather than assigning blame, these case studies are used as learning tools, reinforcing the critical importance of systematic reporting, transparency, and knowledge transfer as foundations for effective accident prevention.

Building on this historical and experiential background, Chapter 3 introduces the fundamental chemical and physical principles governing dust explosion phenomena. Unlike gas or vapor explosions, dust explosions involve heterogeneous systems in which solid particles interact dynamically with the surrounding gas phase. The chapter discusses the key factors influencing dust explosibility, including particle size and morphology, chemical composition, surface properties, oxygen concentration, temperature, pressure, turbulence, and the presence of inert materials or hybrid mixtures. Particular attention is devoted to combustion mechanisms, distinguishing between homogeneous combustion (typical of organic dusts) and heterogeneous mechanisms (characteristic of metallic powders). Simplified modeling approaches for single-particle combustion are also discussed, together with dimensionless analysis tools, providing a theoretical framework that supports both qualitative understanding and quantitative risk assessment.

Chapter 4 translates the fundamental concepts illustrated in Chapter 3 into measurable quantities by focusing on the experimental characterization of dust explosion hazards. The chapter systematically defines and discusses the standardized

DOI: 10.1201/9781003645320-1

parameters used to describe explosion severity and ignition sensitivity, including maximum explosion pressure (P_{max}), maximum rate of pressure rise ($(dP/dt)_{max}$), deflagration index (K_{st}), Minimum Ignition Energy (MIE), Minimum Explosible Concentration (MEC), Limiting Oxygen Concentration (LOC), Minimum Ignition Temperature (MIT), and Layer Ignition Temperature (LIT). Principles, capabilities, and limitations of commonly used experimental apparatuses (such as 20-L sphere, 1 m^3 vessel, Hartmann tube, MIKE apparatus, and ignition temperature furnaces) are reviewed in detail. Emphasis is also placed on both the correct interpretation of laboratory data and the inherent challenges associated with extrapolating standardized test results to complex industrial scenarios.

Recognizing that hazard identification alone is insufficient to ensure safety, Chapter 5 addresses dust explosion mitigation and risk management within existing regulatory frameworks. The chapter provides a structured overview of international, European, and North American standards, including ISO/IEC standards, ATEX directives, and the NFPA regulatory system, with particular attention to the recent consolidation of US requirements under NFPA 660. Beyond regulatory compliance, the chapter introduces risk management as a continuous, systematic process encompassing risk assessment, risk control, and periodic review. Tools such as Dust Hazard Analysis (DHA) are discussed as central elements of a modern, process-oriented safety strategy. The role of human factors, risk perception, and risk communication is also explicitly addressed, acknowledging that technical measures must be complemented by either organizational or behavioral considerations to be truly effective.

Chapter 6 shifts the focus to the practical realities of industrial operations by examining common equipment and processes in the pharmaceutical and food industries. Unit operations such as storage, milling, mixing, drying, granulation, conveying, and dust collection are analyzed with respect to their potential to generate combustible dust atmospheres and, therefore, dust explosions. By explicitly linking equipment-level hazards to material properties such as MIE, K_{st}, and MEC, the chapter provides process engineers and safety professionals with a framework for integrating laboratory data into equipment selection and design, risk assessment, and mitigation strategies choices. This equipment-oriented perspective underscores the necessity of integrating both material behavior and process conditions when addressing dust explosion risks.

To demonstrate how this equipment-oriented perspective can be systematically applied, Chapter 7 presents Quantitative Risk Assessment (QRA) methodologies tailored to dust explosion scenarios in food and pharmaceutical plants. After reviewing classical tools such as HazOp, FMEA, and Fault Tree Analysis, the chapter introduces integrated approaches (most notably the Recursive Operability Analysis Plus with Incidental Sequence Diagrams). Through two detailed industrial case studies, that is, a sugar refinery and a pharmaceutical fluid-bed granulator, the chapter illustrates how probabilistic methods can be used to identify critical accident sequences, quantify explosion probabilities, and evaluate the effectiveness of preventive and protective measures.

Finally, Chapter 8 looks toward the "future" of dust explosion safety by examining emerging challenges and opportunities. The chapter discusses potential evolutions in regulatory practice, including improved standardization of dust explosivity data in safety data sheets and broader adoption of DHA in European legislation. It also explores the growing role of digitalization, artificial intelligence, machine learning, and the Industrial Internet of Things in process safety, highlighting their potential to enhance monitoring, prediction, and decision-making. Advances in mathematical modeling, nanopowder safety, hybrid mixtures, and scale-up methodologies are presented as key research directions for improving both scientific understanding and industrial safety performance.

In summary, this book aims to bridge some of the existing gaps among theory, experimentation, regulation, and industrial practice in the field of dust explosion safety. By combining lessons from past accidents, rigorous scientific foundations, standardized testing methodologies, risk management principles, and forward-looking perspectives, it seeks to provide engineers, safety professionals, researchers, and regulators with a coherent and practical reference for understanding, assessing, and ultimately preventing dust explosion hazards in the pharmaceutical and food industries.

2 Notable Accidents in Pharmaceutical and Food Industries

Given the information currently available, it is both possible and necessary to retrace a timeline of accidental events triggered by dust explosions in the pharmaceutical and food industries. The aim of this section is not merely to report facts, but to encourage reflection among professionals and workers who may, at some point, find themselves involved in activities where dust explosion risks are present. The hope is that those directly responsible for industrial safety can approach their roles with a broader awareness and a deeper respect for the devastating nature of such events and their lasting consequences. Looking back at the past allows us to both better understand the present and imagine a safer future. The events recounted in this chapter are not just tragic episodes confined to their time and place but rather lessons that call us to build stronger foundations for protecting either people or the environment. The intent is that these experiences serve not only as warnings, but also as starting points for renewed commitment and progress. This chapter should also be intended to serve as a reminder: the tragedies that follow had severe human consequences, which often extended far beyond the individuals directly involved. They also brought professional and economic disruption. For all these reasons, doing everything in our power to prevent such accidents from happening again is not just a duty; it is an imperative. The case histories presented here are the result of a thorough bibliographic and documentary research, carefully selected to bring to light the most significant aspects of each accident. The purpose is to convey, from the outset, the destructive power of dust explosions and to show how complete and well-maintained reporting can become a powerful tool for analysis and learning, essential steps toward improving process safety. Crucially, the investigation of past events must not be guided by the search for either blame or identification of a culprit. Such intentions risk distorting the facts, altering details, or even reshaping testimonies in a way that steers responsibility toward one party or another. Rather, the true value of these analyses (and of the reports that follow) lies in their capacity to reflect the reality of events as faithfully as possible, and to draw from them lessons that can help prevent repetition. This is how we can save lives, protect the environment, and preserve resources [1]. Learning from the past, through testimony and structured reporting, means transforming error

DOI: 10.1201/9781003645320-2

into knowledge. It means initiating the learning processes that give meaning, however painful, to catastrophic events such as industrial dust explosions.

2.1 RELEVANT DUST EXPLOSION ACCIDENTS IN PHARMACEUTICAL INDUSTRIES

With regard to the most representative cases of dust explosions in the pharmaceutical sector, it is relevant to present the events summarized in Tables 2.1 and 2.2 which are described in further detail in the dedicated section.

2.1.1 Dust Explosion in West Pharmaceutical Services, Inc. Analysis, Comments, and Learned Lessons

The industrial accident that occurred at West Pharmaceutical on January 29, 2003, resulted in the deaths of six workers and injuries to thirty-eight people, including

TABLE 2.1
West Pharmaceutical Services, Inc. Dust Explosion [2]

Event	Dust Explosions
Company	West Pharmaceutical Services, Inc.
Place and date	Kinston, North Carolina, January 29, 2003
Report type	Official (CSB)
Processing	Production of rubber components for drug delivery, including syringe plungers, vial stoppers, and seals. The manufacturing processes involved rubber mixing, molding, and extrusion
Major damages	6 fatalities, 38 injuries, and significant damage to industrial facilities
Event description	Given the magnitude of the disaster, it is not possible to precisely determine the cause of the explosion or the sequence of events that led to the dispersion and ignition of the dust. However, it has been confirmed that the explosion was fueled by polyethylene dust accumulated in a suspended ceiling
Initiating events	Given the difficulty in identifying the initiating event of the explosion that destroyed the Kinston plant, several scenarios have been hypothesized, such as: 1. Deflagration of vapors released by the decomposition of rubber 2. Dust ignition due to equipment overheating 3. Dust ignition caused by an electric discharge 4. Dust ignition within the engine cooling system
Causes	The analysis conducted on the accident revealed that a combination of multiple factors contributed to the disastrous event. Among these were: 1. Inadequate engineering assessment of the use of zinc stearate and polyethylene in the processing operations 2. Failure to apply certain fire safety standards 3. Failure to identify combustible dust among the materials used 4. A lack of training and information provided to workers regarding the hazards of dust explosions

TABLE 2.2
VNAI Dust Explosion [3]

Event	Dust Explosion
Company	Venkata Narayana Active Ingredients Pvt Ltd (VNAI)
Place and date	Chandrapadia, Andhra Pradesh, India, July 29, 2020
Report type	Unofficial
Processing	Manufacturer of unique Active Pharmaceutical Ingredients (APIs)
Major damages	3 fatalities and 1 injury resulting in permanent disability
Event description	At the time of the explosion, a solution of Allyl-isopropyl-acetyl-urea (AIPAU) powder in methanol at room temperature was being carried out, in preparation for a subsequent recrystallization of AIPAU
Initiating events	It can be assumed that the catastrophic event that affected the VNAI facility was triggered by one of the following three causes: 1. Methanol explosion 2. Dust explosion (AIPAU) 3. Explosion of a hybrid dust–gas mixture
Causes	NA

two firefighters. It is likely the most severe documented case of a dust explosion in a pharmaceutical facility. The explosive event could not be definitively attributed to a specific cause, as the extent of the disaster made it impossible to trace the origin of the explosion. However, investigators agreed that the presence of a polyethylene dust layer in the suspended ceiling significantly contributed to the conditions that enabled the explosion. Other possible causes were analyzed and assessed but ultimately deemed unlikely. As no solid evidence was found to suggest that other materials or conditions could have triggered an explosion of such magnitude, the responsibility for the disaster at West Pharmaceutical was attributed to the accumulation of polyethylene dust in the facility's ceiling structure.

The explosion occurred on January 29, 2003. At that time, normal operations were reportedly underway, and none of the witnesses interviewed by investigators recalled any signs that might have indicated the imminent event. At 1:28 PM, operators heard a loud bang followed by the collapse of part of the building where the compounding process was taking place, accompanied by a fireball and a rising cloud of smoke. The entire facility was compromised by the explosion, with the most severely affected areas being the compounding unit (where the explosion originated) and the milling department (where the blast wave was most intense). Both the outer cladding and the walls of the compounding area were destroyed, the warehouse collapsed, and those parts of the building that remained standing were declared unsafe. The explosion and resulting fire not only affected the entire structure but also disabled the emergency systems and sprinkler lines. The most persistent fire broke out in the warehouse and burned for two days before being extinguished.

Following the event, investigations were launched to either determine what had occurred or draw lessons from the tragedy. Based on the collected information and photographic evidence, inspectors reconstructed the cleanliness conditions of the milling area. It was found that dust accumulation near the machinery was minimal, thanks to the constant intervention of cleaning crews. However, a substantial layer of polyethylene dust was present in the suspended ceiling, especially in the area directly above the machinery. This accumulation was due to certain machines emitting fine dust as a process residue (classified as fugitive emissions). The presence of an air conditioning system in the ceiling area created a slight negative pressure, which drew airborne particles upward, facilitating their deposition on the ceiling surface. In such conditions, even experienced professionals or inspectors would have struggled to detect the presence of dust, as the accumulation occurred in an area that was neither visible nor included in cleaning procedures. During several external inspections of the site prior to the explosion, neither inspectors nor industrial hygienists identified this issue. The work area beneath the dust accumulation zone was regularly cleaned, and even the safety personnel were unaware that the dust had combustible and explosive properties. The suspended ceiling also housed lines for conveying calcined clay and other non-combustible powders, suggesting that the accumulated material likely consisted of a mixture of polyethylene dust, calcined clay, and other non-combustible substances. These inert powders may have partially inhibited the explosive behavior of the polyethylene.

Nevertheless, as no other material present at West Pharmaceutical could have generated an explosion of such magnitude, investigators identified the polyethylene dust accumulation as one of the key contributing factors to the disaster. What remains unclear is what caused the accumulated dust to become airborne, as a dust explosion requires dispersion into the air to occur. On this point, the investigation did not reach a conclusive determination. However, investigators agreed that the presence of suspended dust was a critical element and that dust removal is one of the most important preventive measures since it is not feasible to eliminate all potential ignition sources in an industrial setting. The extreme difficulty in identifying what lifted the dust into suspension, as well as the precise ignition source, is largely due to the catastrophic consequences of the event itself. The violent explosion and prolonged fire destroyed much of the physical evidence that could have supported a thorough post-incident analysis. Overall, regardless of the exact ignition source, investigators agreed that the greatest hazard was the presence of accumulated dust, whereas the type of ignition source played a secondary role.

Although dust accumulation was identified as the fundamental factor in the explosion, the investigators who visited the facility after the event quickly realized that the general understanding and perception of dust explosion risk had been very poor, to the extent that the significance of such accumulations was underestimated. This lack of attention extended even to inspectors who had visited the plant in the months prior to the explosion. A summary of the conclusions drawn in some of the reports issued to the facility is provided below:

- An industrial hygienist assessed worker exposure to airborne dust but made no reference to its explosiveness.

- An official inspection by NC-OSHA (North Carolina Department of Labor's Occupational Safety and Health Division) took place just months before the explosion and did not identify the presence of dust accumulations.
- A risk insurance company visited the facility and, although recommending the removal of dust from sprinklers, made no mention of the dust explosive properties.

The explosion at West Pharmaceutical prompted authorities to conduct a detailed investigation in order to identify its causes and prevent similar conditions from arising in other industrial contexts. It is, therefore, essential to establish a robust post-accident analysis and review system, not for the purpose of assigning blame but to enable other facilities to protect themselves from such scenarios, and to allow academic institutions to examine these events, share knowledge, and contribute to the ongoing improvement of process safety in increasingly complex industrial environments.

Such well-structured investigations and the documentation that follows are key to reconstructing the event and deriving critical lessons regarding behaviors and practices that can protect workers lives. This case has clearly demonstrated how the absence of appropriate measures and precautions to prevent dust explosions can significantly increase the likelihood of a disaster.

2.1.2 Dust Explosion at VNAI Pvt Ltd

On July 29, 2020, three people lost their lives and another suffered severe and permanent injuries following an explosion at the VNAI Pvt Ltd factory in Chandrapadia, a city in Andhra Pradesh, India. The information available about this tragic event comes from unofficial sources [3], which themselves rely on confidential official documents obtained anonymously.

According to a recent market analysis article, it is worth highlighting the steady international growth of the pharmaceutical industry in India. In 2023, there were officially 10,000 pharmaceutical manufacturing plants registered in India, of which 650 comply with FDA (Food and Drug Administration) standards, the highest number of FDA-compliant facilities outside the United States. This makes India the largest supplier of generic drugs present in the world. The global contribution of the Indian pharmaceutical industry stands at approximately 3.5%, with future projections estimating a rise to a 5% share of the world market by 2030, corresponding to an economic value of roughly 120 to 130 billion USD. Long-term estimates predict the Indian pharmaceutical market to reach about 450 billion USD by 2047 [4]. This notable growth, both current and projected, should be viewed in conjunction with considerations regarding the existing industrial safety regulatory framework, which may still have room for further development.

From a technical standpoint, it is important to note that official reports concerning this event provide only sparse information about the accident analysis. Given this scarcity of data, researchers conducting the unofficial investigation were compelled to undertake independent studies and research to identify the most likely cause of the explosion. Their investigation resulted in the identification of three potential initiating events, each examined in depth to determine the most credible one. According to available information, at the time of the explosion, a process was underway involving the dissolution of AIPAU powder, also known as Apronal, in methanol at room temperature, in preparation for a subsequent recrystallization of the powder.

Regarding this specific process, the identified possible explosive scenarios are:

1. Explosion of methanol vapors
2. Explosion of AIPAU dust
3. Explosion of a hybrid dust–vapor mixture

2.1.2.1 Explosion of Methanol Vapors

The independent study focused on this scenario 1 began with an analysis of the preliminary and final reports available to the researchers. These reports state:

- "At about 02:10 AM, static electricity was generated and developed a spark, which ignited methanol vapors accumulated in that room and caused the fire accident, resulting in damage to the walls of the room, and all four workers present suffered severe burn injuries."
- "It is learnt during the course of preliminary enquiry that at about 2:10 AM on 07.29.2020, methanol vapours accumulated in clean room-2 caught fire, presumed to be due to static electricity generation in the SSR/120 Reactor and a spark developed while charging the chemical powder of AIPAU into the reaction vessel through its manhole."
- "Since the management adopted the procedure for charging chemical powder into the Reactor through its manhole, it allowed a large quantity of methanol vapors and powder dust generated inside the vessel during the process to escape into the clean room, causing the accumulation and formation of a vapor and dust cloud, which is an unsafe practice."

The dissolution of AIPAU in methanol was carried out at room temperature, and the unofficial analysis considered a methanol vapor pressure value of 20 mmHg at approximately 30°C. However, this value is underestimated compared to those widely accepted in the scientific community, where vapor pressure values around 150 mmHg at 30°C are reported [5]. This underestimation consequently characterizes the substance as less hazardous than it actually is, in terms of reaching flammability conditions. In conclusion, the official report obtained by researchers Babu Rao, Rao, and Venkat Reddy states that "At such low vapor pressure, methanol vapors accumulating in the clean room is not probable."

2.1.2.2 AIPAU Dust Explosion

AIPAU powder (Apronal) is classified as non-flammable [6], although the report indicates it as flammable. Similar to the contradiction observed regarding the methanol vapor pressure, this inconsistency supports the view that it would be highly desirable to include every country involved in industrial production within a system for drafting, reviewing, and disseminating official documents concerning production events and industrial disasters of international relevance. According to the unofficial report, the investigating commission "has not examined the possibility of AIPAU dust explosion." Therefore, this scenario has not been sufficiently analyzed to reach a conclusion on its significance, whether major or minor, in triggering the explosion.

2.1.2.3 Hybrid Explosion

Based on report [3], it is plausible to assume that the explosion at the VNAI Pvt Ltd factory in Chandrapadia was caused by a hybrid mixture explosion, where a solvent (methanol) combined with a powder (AIPAU or Apronal), facilitating ignition compared to powder alone and increasing the explosive force. In the case study analysis, it is hypothesized that while charging the ninth bag of AIPAU powder, mechanical friction between the falling particles was sufficient to create a flash fire, given the hybrid mixture of dust and solvent vapors present in the vessel. Additionally, the crude AIPAU is likely to contain a solvent like xylene from Stage IV of production, though its composition is not provided. With the manhole approximately 450 mm in diameter open (no vessel data available), the primary explosion/deflagration caused violent dispersal of AIPAU dust into the clean room through the open manhole, followed by a more severe secondary explosion. According to the final enquiry report, supervisors rushing to clean room-2 after the explosion found "Mr. Sayyad Hafeez standing at the spot with severe burn injuries and his body covered with chemical powder."

2.1.3 Other Explosion Accidents in Pharmaceutical Industries

Tables 2.3–2.6 briefly illustrate further significant accidents having affected the pharmaceutical industry.

TABLE 2.3
Auribondo Pharma Explosion [7]

Place and date	Andhra Pradesh, Srikakulam (India), August 11, 2019
Activity involved	Aurobindo Pharma, Pharmaceutical Manufacturing
Brief description	Inside the plant, a boiler had become blocked by ash dust, causing an explosion due to overpressure. The event resulted in the direct death of two people and serious injuries to a third. The company had already been affected by four accidents (between 2011 and 2014), resulting in a total of 10 injuries and two deaths

TABLE 2.4
Tianmen Chutian Fine Chemicals Company [8]

Place and date	Tianmen City, Hubei (China), September 28, 2020
Activity involved	Tianmen Chutian Fine Chemicals Company, Pharmaceutical Compounding
Brief description	The accident was believed to have been caused by a powdered chemical that, after exploding, released yellow smoke into the atmosphere, indicative of the presence of nitric acid. The event resulted in the deaths of five people and injuries to one other

TABLE 2.5
Shri Nivas Gujarat Laboratory Explosion [9]

Place and date	Baddi, Himachal Pradesh (India), October 29, 2021
Activity involved	Shri Nivas Gujarat Laboratory, Pharmaceutical Manufacturing
Brief description	The explosion occurred due to a short circuit in the tablet drying system. The event and the subsequent fire injured five people

TABLE 2.6
Sigachi Industries Explosion [10]

Place and date	Sangareddy, Telangana (India), June 30, 2025
Activity involved	Sigachi Industries pharmaceutical, Pharmaceutical Manufacturing
Brief description	Preliminary reports suggest the explosion was caused by a massive increase in temperature (700–800°C) inside the spray dryer during the production of microcrystalline cellulose. The consequences of the event resulted in the deaths of 40 people and the injuries of at least 33 others

2.2 RELEVANT DUST EXPLOSION ACCIDENTS IN FOOD INDUSTRIES

Regarding the most representative cases of dust explosions in the food industry, it is useful to refer to the events summarized in Tables 2.7 and 2.8 which are described in greater detail in the dedicated sections.

2.2.1 Dust Explosion in Malt House Silo (Metz, France). Analysis, Comments, and Learned Lessons

Regarding a possible and hypothetical currency exchange, the damages caused solely to the company's structural facilities (due to the explosion at the malt processing

TABLE 2.7
Malt House Silo, Dust Explosion [11]

Event	Dust Explosion
Company	Non-Officially Declared
Place and date	Metz (Moselle) – France, October 18, 1982
Report type	Official (ARIA)
Processing	Barley malting process for use in the brewing industry
Major damages	12 deaths, 3 injured (including one seriously), significant damage to company facilities, and contamination of the Moselle River alluvial fan
Event description	The primary explosion occurred inside the malt processing tower, creating conditions that triggered two secondary explosions, which ultimately caused a significant portion of the structure to collapse
Initiating events	A possible cause of ignition appears to have been the carelessness of an operator who was smoking near the processing tower, in an environment where barley and malt dust were present
Causes	Overall, the risk of fire and explosion had been significantly underestimated by the entire workforce. Several specific practices contributed to the occurrence of the incident: 1. No safety documentation was ever shared for the production site. 2. No procedures were ever established for internal and subordinate staff regarding precautions to take during both routine and non-routine operations. 3. No explicit smoking ban was ever enforced inside or near the facilities. 4. The installed equipment lacked pressure relief systems to vent overpressure in the event of an explosion.

site) amount to a current equivalent of about € 27,000,000 [13], which, although it excludes other types of consequences, is potentially sufficient to cause the permanent closure of the company. Furthermore, as it will be further detailed in Section 2.4, it is important to anticipate that, to obtain a comprehensive estimate of the damages caused by an accidental event, it is necessary to add to the structural damages an estimate of other direct and indirect damages, which significantly increase the economic magnitude of the loss. Based on the application of the European Scale of Industrial Accidents (ESIA), the general extent of the damages caused by this event can be summarized as shown in Figure 2.1.

The explosion that affected the malt processing plant caused the death of thirteen people and injuries to other three, one of whom was seriously injured, and he was rescued alive only a few hours after the collapse of the structure in which he was located. The towers that constituted the storage reserves partially collapsed, and those that remained standing suffered damages sufficient to be considered unstable. After the accident, large quantities of malt and barley residues were removed from the plant area and disposed of in a former gravel quarry in Moulins-les-Metz. This practice led to pollution of the alluvial cone of the Moselle River, and in 1983, by decree of the responsible Prefecture, a remediation cost payment order was issued

TABLE 2.8
Imperial Sugar Company, Dust Explosion [12]

Event	Dust Explosion
Company	Imperial Sugar Company
Place and date	Port Wentworth, Georgia (USA), February 7, 2008
Report type	Official (CSB)
Processing	Processing of raw cane sugar into granulated sugar to produce powdered sugar, specialty sugar, or liquid sugar
Major damages	14 deaths, 36 injuries, and significant damage to company facilities
Event description	The primary explosion that triggered the disastrous event was most likely caused by overheating of a bearing in a closed steel belt conveyor located beneath silos 1 and 2, which contained sugar. This accident led to a series of secondary explosions that caused walls to collapse and floors to rupture. The fires resulting from the explosions then destroyed buildings, structures, and company facilities, causing severe damage as well as multiple fatalities and injuries among workers
Initiating events	A few days before the accident, it was necessary to remove lumps of sugar stuck in the discharge system of silo 1 on the enclosed conveyor belt. The procedure followed caused some of these lumps to become trapped between the conveyor belt and the discharge chute of silo 1, creating an obstruction that blocked the normal flow of sugar along the line. Since the sugar could not pass regularly, it accumulated on the side of the belt. Sugar dust produced by friction and lateral spillage then accumulated in the unventilated area of the conveyor. This buildup, in contact with an ignition source, most likely an overheated the bearing and triggered the primary explosion
Causes	The analysis of the incident revealed that a combination of several factors contributed to the disastrous event. Among these, there were: 1. The sugar transport equipment was neither designed nor maintained adequately to minimize the dispersion of sugar dust into the work area. 2. Inadequate cleaning practices led to the accumulation of sugar dust layers on floors and surfaces. 3. Combustible sugar dust suspended in the air had accumulated beyond the minimum explosive concentration inside the new enclosed steel conveyor system beneath silos 1 and 2. 4. The evacuation plans for the facility were inadequate.

against the malt house amounting to five million francs, approximately € 1,750,000 in current value [13]. The remediation effort involved removing about 12,500 m³ of malt and barley from the former gravel quarry for delivery to an authorized disposal center.

The explosive event occurred around 14:15 on October 18, 1982. At the time of the disaster, approximately 12,000 tons of barley and malt dust at low moisture levels (16% for barley and 4% for malt) were present inside the storage complex. Work activities that afternoon involved subcontractors, boiler operators, and masons engaged in various roles related to the operation and maintenance of the plants and

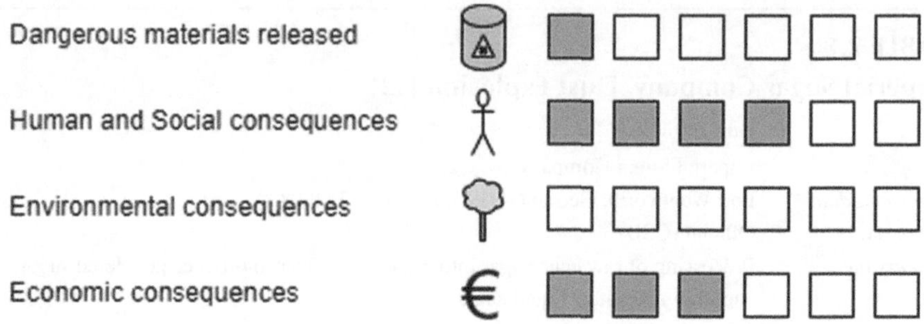

FIGURE 2.1 ESIA scale results applied on dust explosion in malt house silo [11].

structures. When the ignition, probably caused by the carelessness of a smoker involved in work activities, encountered the explosive atmosphere in the storage area, a primary explosion occurred. Witnesses in other parts of the plant reported hearing two explosions of increasing intensity, confirming that the primary explosion generated conditions for a more severe secondary explosion which resulted in tragic consequences.

Although the effects of this event were confined within the facility, the French Minister of Environment deemed them sufficiently serious to initiate a general inspection aimed at clarifying the causes of the explosion and gathering as much information as possible related to the accident. The authorities' intent was to clarify the dynamics of the accident to improve safety in the future management of activities related to storage silos. The final report, drafted on February 7, 1983, unequivocally stressed the need to adopt appropriate cleaning practices to prevent dust accumulation in such installations and to carry out frequent removal of these deposits; furthermore, it emphasized the importance of implementing operational and maintenance procedures to ensure greater worker and process safety. The French Legislative Official Gazette was promptly updated after the October 1982 event with the inclusion of a specific technical memorandum for silos containing food products capable of releasing flammable dust. Nine years after the explosion, the Metz Court of First Instance sentenced the malt house general manager to six months imprisonment and a fine currently equivalent to about € 5,100 for manslaughter and injuries.

The significance of human awareness in relation to the disaster is what stands out most from this dramatic event, where workplace neglect, the absence of smoking bans near explosive atmospheres, and lack of specific safety training were fundamental elements in triggering the event. Human and organizational factors, sometimes considered marginal compared to more understandable technical and technological improvements, created the ideal conditions for the explosion to occur. The presence of a dust-laden environment not removed by cleaning activities and the presence of a cigarette could have been avoided if greater attention had been paid to properly describing the standard operating procedures for both work and maintenance of the facility. It is important to emphasize again that the absence of smoking and combustion bans constituted a fundamental difference in the effectiveness of explosion prevention. Behind this deficiency was the lack of a structured risk analysis, which, fortunately, today is a widely adopted practice as a preventative measure against incidental events. The fact that the malt house management was unprepared and unaware of the risks present in the plant also highlights how the involvement of all workforce levels is necessary to implement safety for everyone.

Finally, the experience and awareness gained from this specific event also teach that proper waste management prevents further unnecessary environmental repercussions that occur when waste disposal is rushed and ambiguously conducted in unsuitable locations. This, beyond leading to a likely significant economic sanction, has the potential to pollute various natural environments accessible to people and animals.

2.2.2 Dust Explosion at Imperial Sugar Company. Analysis, Comments, and Learned Lessons

Granulated sugar from the refinery was stored in three 32-meter-tall silos before being conveyed to packaging buildings for distribution. Additionally, the granulated sugar was processed into specialty products such as brown sugar and powdered sugar. The sugar was transported through a complex system of bucket elevators, screw conveyors, and conveyor belts. During these operations, sugar frequently spilled onto floors across the work areas, accumulating in some places to several inches in depth. This spilled sugar contained fine particles that became airborne, contributing to a hazardous dust environment. Hammer mills were employed to crush granulated sugar into powdered sugar, generating additional dust. Although connected to a dust collection system, the equipment was undersized, poorly maintained, and did not cover the bucket elevators or conveyor belts. Consequently, significant quantities of sugar dust escaped into the workplace. Workers routinely used compressed air to clean packaging machines, further dispersing sugar dust throughout the buildings. Over time, dust accumulated on elevated, hard-to-clean surfaces, such as beams and lighting fixtures, which were not cleaned sufficiently to maintain dust levels below hazardous thresholds. Beneath the sugar silos, granulated sugar flowed through chutes onto a long steel conveyor belt. Occasionally, sugar clumps obstructed one of the chutes, blocking sugar flow on the belt, causing spills, and releasing dust into the tunnel. The tunnel, however, was large and ventilated, preventing the buildup of dust to explosive concentrations.

In 2007, the conveyor belt was enclosed with stainless steel panels to protect the sugar from contamination. This enclosure, however, was not equipped with a dust collection system, resulting in the accumulation of sugar dust within a confined and unventilated space. On February 7, 2008, clumps of sugar blocked one of the discharge chutes, leading to sugar spilling off the moving belt. Dust likely accumulated inside the enclosure to explosive concentrations. Around 7:15 PM, sugar dust contacted an ignition source, possibly an overheated bearing, triggering a primary explosion. This explosion breached part of the enclosure and vented into the adjacent packing building. The lofted sugar dust ignited by advancing fireballs fueled a chain reaction of secondary explosions that swept through the buildings. Concrete floors buckled, releasing tons of granulated and powdered sugar into the flames. Emergency evacuation drills had never been conducted, and the explosions caused widespread power outages, plunging stairwells and passageways into darkness, hindering workers' escape efforts. Subsequent analysis of the Imperial Sugar Company incident highlighted that the Wentworth plant had significantly underestimated the risks posed by sugar dust. Despite awareness of explosion hazards dating back to 1925 and formal requests since 1958 to implement adequate equipment and cleaning protocols, the company failed to prioritize risk mitigation measures. Workers confirmed that minor sugar dust ignitions had occurred in previous years, providing clear evidence of the dust flammability. Moreover, two weeks prior to the catastrophic explosion, a minor blast in a plant collector was effectively contained due to properly sized explosion panels.

Despite this knowledge and prior accidents, management did not fully recognize the severity of the hazard. This is exemplified by the 2007 decision to enclose the conveyor belt without incorporating explosion protection measures. The enclosed space beneath the silos thus became a confined, unventilated area where sugar dust could reach explosive concentrations. Additionally, the enclosure lacked explosion venting panels, preventing safe relief in the case of a blast. In conclusion, a combination of increased awareness, proper investment in equipment, and adherence to rigorous safety procedures could have prevented or significantly mitigated the impact of this tragic event.

2.2.3 Other Explosion Accidents in Food Industries

Tables 2.9–2.12 briefly illustrate further significant accidents affecting the food industry.

2.3 STATISTICS OF DUST EXPLOSIONS

One of the first documented analyses of a powder explosion dates back to 1786. At that time, Count Carlo Ludovico Morozzo di Brianzè presented his "Relation d'une violente détonnation arrivée a Turin le 14 décembre 1785 dans un magasin de farine" to the Academy of Sciences of Turin without knowing that he would lay the foundation for the data collection and documented investigation of powder explosions

TABLE 2.9
Qianhuangdao Lihua Starch Explosion [14]

Place and date	Funing, Qinhuangdao (China), February 25, 2010
Activity involved	Qianhuangdao Lihua Starch Co Ltd
Brief description	The explosion, triggered by cornstarch; it killed 19 people and injured 49 others

TABLE 2.10
Arva Spices Explosion [15]

Place and date	Wijhe, Overijssel (Netherlands), September 11, 2019
Activity involved	Arva Spices, Spice Manufacturing
Brief description	The dust explosion was caused by flames escaping from a spice grinding machine. A facade of the building was demolished by the blast wave, and three people were injured in total

TABLE 2.11
Parry Sugar Explosion [16]

Place and date	Kakinada, Andhra Pradesh (India), August 19, 2022
Activity involved	Parry Sugars Refinery, Sugar Production
Brief description	The sugar dust was ignited by an exposed electrical cable, which caused an explosion along the conveyor belt where the sugar is loaded. The accident occurred when the power supply was activated to start the conveyor belt loading the products onto the truck. The accident resulted in two deaths and nine injuries

TABLE 2.12
Oba Makarna Explosion [17]

Place and date	Hendek, Sakarya (Turkey), September 15, 2025
Activity involved	Oba Makarna, Pasta factory
Brief description	A short circuit in the electrical panel ignited grain dust in the milling section of the plant. The main causes of this event may have been incorrect risk analyses, layered dust accumulation, and inadequate electrical systems. The event resulted in 5 deaths and 30 injuries

in process activities [18]. However, it is only in more recent times that the practice of systematically recording and studying accidents caused by dust explosions has become consolidated, allowing the collection of significant amounts of data with which it is possible to make comparisons, trace temporal trends, and identify possible correlations even in apparently different contexts [19].

2.3.1 From 1800

The first information and data available on dust explosions in industrial environments indicate that in the period between 1800 and 2010, almost half of the recorded accidents occurred in the United States (1161 events out of 2420) and that, among those that occurred in Europe, the majority concerned both Germany (426 events) and the United Kingdom (411 events). The data for each country are reported in Figure 2.2 and, given the long period analyzed, it is important to consider how industrial development, both global and specific, may have influenced the total number of accidents recorded, given the closed link between work activities and dust explosions [20].

It is important to underline that the number of dust explosion events, both industrial and non-industrial, has decreased over time, as highlighted in Figure 2.3, also thanks to the improvement of the available safety devices and systems. In contrast to this phenomenon, China, as reported in Figure 2.4 has been affected by an increasing

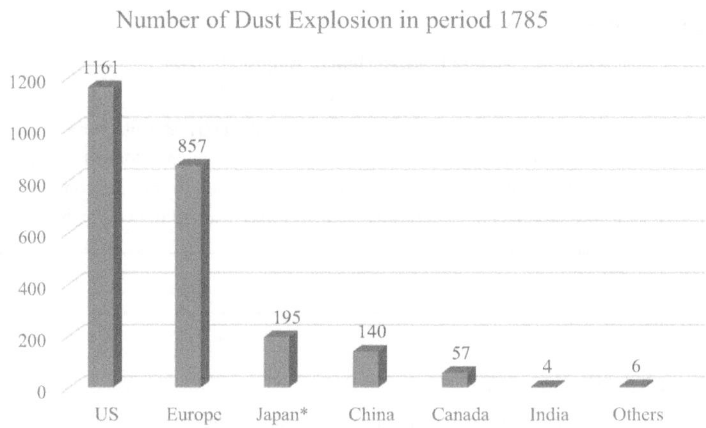

FIGURE 2.2 Dust explosions occurred in the period 1785–2012 [20].

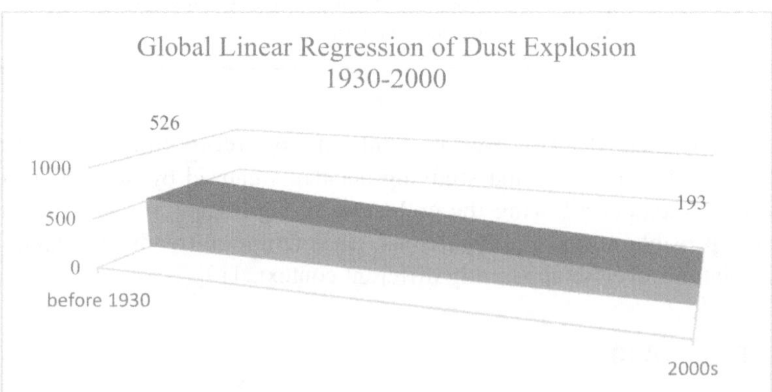

FIGURE 2.3 Linear regression of the number of dust explosions recorded globally between 1930 and 2000.

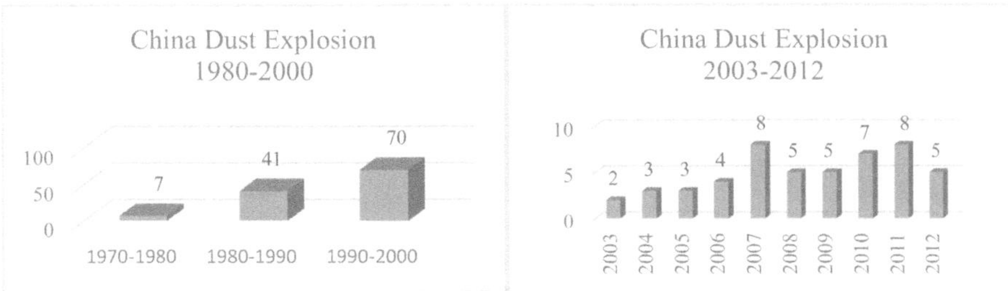

FIGURE 2.4 Number of dust explosion in China considering different period of times [20].

TABLE 2.13
Details of Dust Explosions in the US Food Industry, 1900–1956 [21]

Territory	Type of Industry	Period	Explosions	Damage (in $)	Deaths	Injuries
US	Food and Feed	1900–1956	577 (51%)	75.8 million	409 (61%)	1061 (60%)

number of dust explosions probably associated with the industrial development of the last decades.

The food industry is notoriously among the most sensitive to dust explosions and the United States has experienced this as reported in Table 2.13.

Approximately half of the events that occurred in the period 1990–1956 in the USA involved food processing plants, causing significant damages, estimated at approximately $76 million, and the death or injury of more than half (about 60%) of the people involved in dust explosions in the period indicated [21]. In the United States, the Chemical Safety Board (CSB) has compiled a report containing information on dust explosions, an extract of which covering the period 1980–2005 can be analyzed. For each year, events relating only to the food and pharmaceutical industries were selected, and the temporal trend of the recorded events is shown in Figure 2.5.

The conducted analysis considered some specific elements regarding both the type of activity and the type of fuel involved, as reported in Table 2.14.

The events considered by CSB highlight how:

- The chemical-pharmaceutical sector recorded a total of six explosions between 1980 and 2005. In 83.3% of cases, the fuel was a pharmaceutical dust, while in the remaining 16.6%, it was an inorganic dust.
- The food production and packaging/storage sector recorded a total of 68 explosions between 1980 and 2005. In 2.94% of cases, the fuel was wood dust, while in the remaining 97.06% of cases, it was food dust.

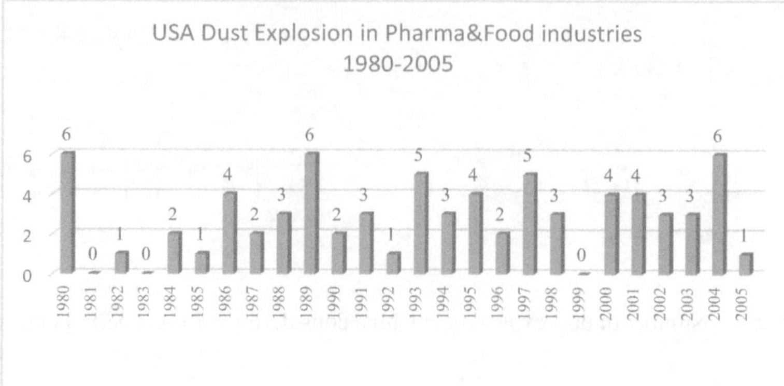

FIGURE 2.5 Number of dust explosions in food and pharmaceutical industries in the USA (1980–2005).

TABLE 2.14
Reference Elements for Dust Explosion Analysis in USA (1980–2005)

Type of activity	Food production
	Chemical-pharmaceutical production
	Packaging/storage
Type of combustible	Food
	Pharmaceutical
	Inorganic
	Wood

TABLE 2.15
Details of Dust Explosion Analysis in the US Food and Pharmaceutical Industries (1980–2005)

Activity	Explosion	Fatalities	Injuries	Fatalities/Explosion	Injuries/Explosion
Pharmaceutical	6	3	7	0.5	1.12
Food	68	12	106	0.18	1.56

It is also necessary to emphasize that, in the analysis referring to the food industry, two events that originated from a dust not typically associated with the food industry were deliberately included. The presence of wood dust in workplaces represents a major risk due to the potential and high capacity of this fuel to develop a catastrophic explosion. Therefore, although it is not a food dust, the decision to include it in the analysis aims to show how an unusual element can lead to accidental events in different types of companies, thereby stimulating critical thinking during risk analysis activities. Table 2.15 presents the summary data of this investigation, from which it

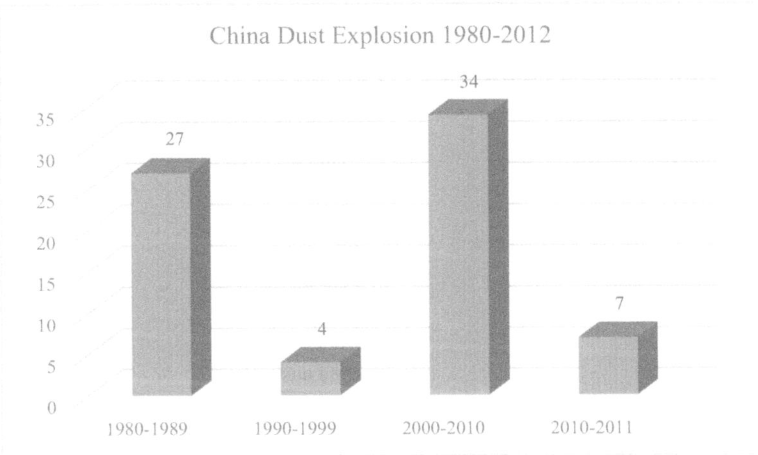

FIGURE 2.6 Number of dust explosion in China [23].

is also possible to observe how explosions in the pharmaceutical industry, although fewer in numerical terms, are capable of causing a greater number of fatalities and injuries compared to those related to the food industry [22].

Focusing now on the Chinese panorama, it is interesting to see how a different source compared to the previous one makes available additional data to those reported in Figure 2.4. Within Figure 2.6, it is, therefore, possible to see the trend of dust explosions recorded in China in the decades from 1980 to 2011.

The recorded explosions refer to various types of facilities and work activities, excluding those dedicated to the processing of spices, sugar, cocoa, and pharmaceuticals, as these industries, at the time of the analysis, were not yet significantly widespread in China. The focus was, therefore, on explosions that occurred in the food industry, unlike what was done for the United States, where the analysis was conducted by selecting only a few specific elements, as explained in Table 2.16 [23].

As previously indicated in Table 2.14, Table 2.16 also includes a dust not typically associated with explosion risks in a food industry, namely ash (explosive event that occurred in 1986 and resulted in injury to one person). In line with what has already been said, we reiterate our intention to promote risk analyses that can include elements that are sometimes unusual for certain types of work activities.

In relation to the available data, it is possible to report what is summarized below and reported in Table 2.17.

- For the types of industries reported in Table 2.16 present in the Chinese territory during the investigation period, a total of 31 dust explosions occurred, 4 for the starch/candy sector (12.9%) and 27 for the flour/feed sector (87.1%).
- The overall consequences of these explosions resulted in the deaths of 29 people and the injury of 195 others. The most catastrophic event occurred in 2010 due to the ignition of starch dust. The explosion caused 19 deaths (65.5% of the total) and 49 injuries (22.1% of the total).

TABLE 2.16
Reference Elements for Dust Explosion Analysis in China (1980–2011)

Kind of activity	Flour and Feed
	Starch and Candy
Kind of combustible	Food dust
	Wheat flour
	Feed dust
	Ash
	Bran dust
	Corn dust
	Malt dust
	Starch

TABLE 2.17
Details of Dust Explosion Analysis in Food Processing Industry in China (1980–2011)

Activity	Explosion	Fatalities	Injuries	Fatalities/Explosion	Injuries/Explosion
Flour/feed	27	10	99	0.37	3.66
Starch/candy	4	19	96	4.75	23.5

2.3.2 Recent Years (2016–2023)

For recent periods, more organized data collection is available. For the period 2016–2023, the number of dust explosions recorded worldwide, with details for the USA and Canada, is shown in the graph of Figure 2.7.

Similar to data in Figure 2.7, Figure 2.8 shows the trend in dust fire events that occurred in the period 2017–2023.

As it can be seen from the numerical comparison shown in Figure 2.9, the number of accidents associated with dust fires is significantly higher than that of explosive events. This aspect is relevant for process safety because, even in the absence of confinement, an accident can occur that can cause significant losses: human, environmental, and company capital.

Dust fires and dust explosions have resulted in significant losses of life and injuries in recent years. In the period considered (2016–2023), a total of 231 victims and 902 casualties were recorded, which could be potentially underestimated values compared to the real data. Two particularly significant events impact the estimates and statistics that can be made on the basis of this data: in Russia (2021) and Turkey (2022), two dust explosions occurred in mines, resulting in 51 and 41 deaths

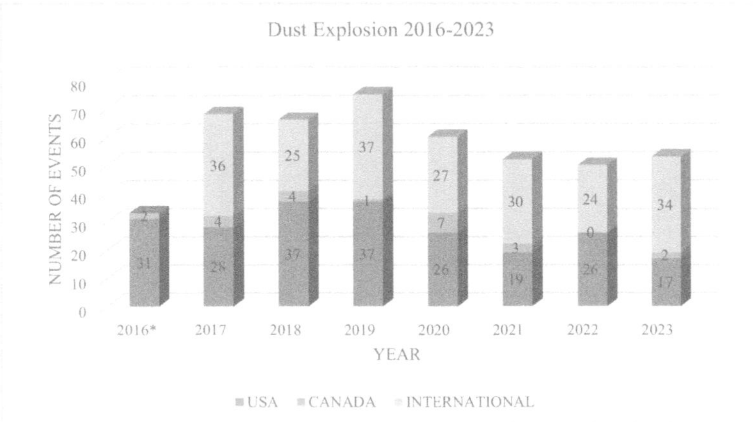

FIGURE 2.7 Dust Explosion from 2016 to 2023 [24–29].

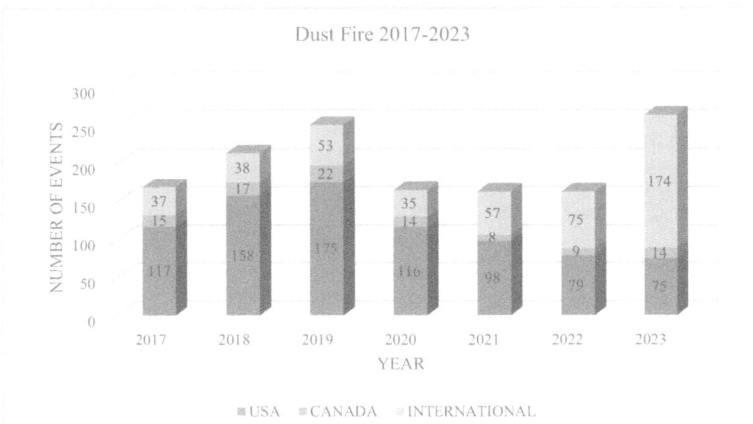

FIGURE 2.8 Dust fire from 2017 to 2023 [24–29].

respectively, in addition to injuring 106 and 11 people [27, 28]. A more detailed observation of how these losses are distributed over time is reported in Figure 2.10.

The data reported in Figure 2.10 refer to the consequences caused by both explosions and fires due to dust ignition. A more precise distribution of the relationship between consequences and event type, referring to the period 2018–2023, is reported in Tables 2.18 and 2.19.

By developing an average over the time period for which the data are available, a statistical distribution of deaths and injuries is obtained in relation to the single type of event which is reported in Figure 2.11.

During this period, food and pharmaceutical industries have been involved in several accidents caused by dust ignition, as reported in Table 2.20.

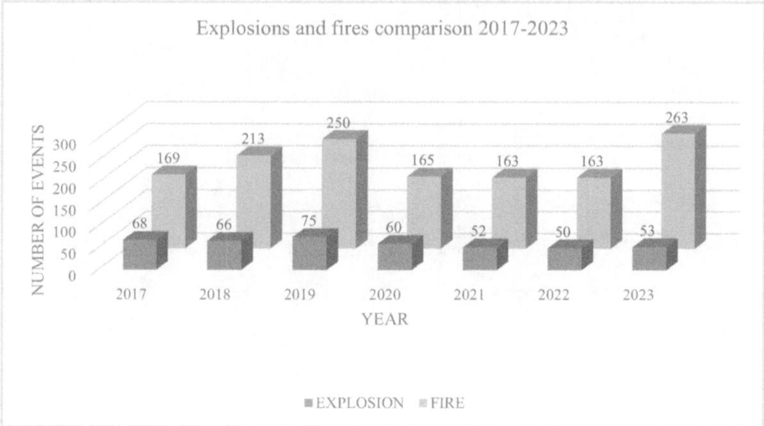

FIGURE 2.9 Dust explosion and dust fire comparison 2017–2023 [24–29].

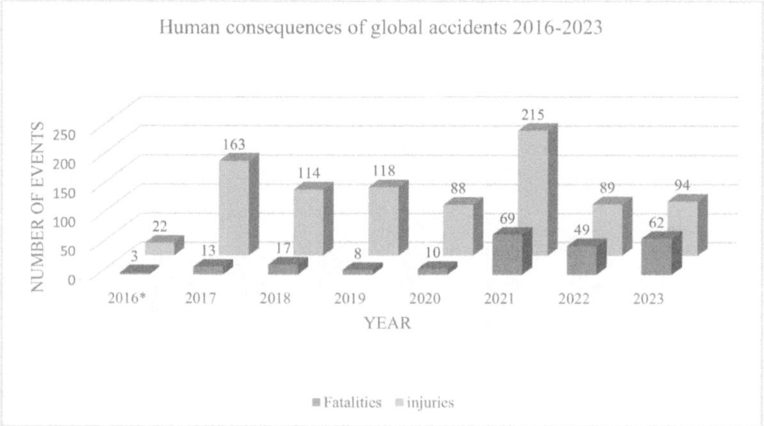

FIGURE 2.10 Human consequences of global accidents 2016–2023 [24–29]. *Data available for USA and Canada only.

TABLE 2.18
Relation between Kind of Event and Number of Injuries [24–29]

Year	Total Number of Injuries	Dust Fire (%)	Dust Explosion (%)
2018	114	16	84
2019	118	27	73
2020	88	27	73
2021	215	17	83
2022	89	16	84
2023	94	14	86

TABLE 2.19
Relation between Kind of Event and Number of Fatalities [24–29]

Year	Total Number of Fatalities	Dust Fire *(%)*	Dust Explosion *(%)*
2018	17	6	94
2019	8	13	87
2020	10	30	70
2021	69	8	92
2022	49	57	43
2023	62	34	66

FIGURE 2.11 Average of injuries and fatalities with respect to the single kind of event for 2018–2023 [24–29].

Figure 2.12 provides a detailed description of the relationship between fires and dust explosions for the activities most relevant to the purpose of this text, especially given the significant presence of food dust within the work areas.

From data reported in Table 2.20, it can be seen that wood, food processing, and agricultural industries are the ones most affected by accidents caused by the ignition of combustible dusts. Table 2.21 instead reports the types of fuels most involved in dust fires (F) and dust explosions (E), with reference to the number of events that were determined by the specific fuel for each year analyzed.

The types of fuel involved in the explosions that affected different work sectors are shown in the graphs of Figures 2.13– 2.15.

It is evident that grain or grain dust were responsible for the majority of the recorded accidents (approximately 68.5% of the cases recorded in the agricultural, food processing, and ethanol production sectors). However, it is necessary to reiterate that in these sectors there is also a risk associated with non-food fuels, such as wood dust or sawdust (approximately 1.8%) or chemical dust (approximately 0.3%). A residual percentage of accidents, approximately 2%, was caused by the combustion of unrecognized materials. A total of 67 events were caused by fuels classified as "others," which includes the fuels listed in Table 2.22.

TABLE 2.20
Activities Involved in Dust Fire or Dust Explosions [24–29]

Year	Agric and Food (%)	Food Process (%)	Ethanol Production (%)	Wood Industry (%)	Automotive and Metal Industry (%)	Power Industry (%)	Mining (%)	Others* (%)
2016	33.0	No data	No data	33.0	9.0	9.0	No data	16.0
2017	33.0	No data	No data	28.0	11.0	No data	No data	29.0
2018	26.0	14.5	2.5	24.0	11.0	4.0	2.0	17.0
2019	34.7	8.2	0.9	21.0	9.0	0.3	1.0	25.0
2020	32.9	17.3	0.9	22.7	8.0	1.8	0.4	16.0
2021	36.5	11.6	2.3	18.5	5.1	4.6	1.9	19.5
2022	38.0	9.4	2.4	20.2	10.8	1.4	0.0	17.8
2023	31.6	12.2	0.3	28.1	6.0	2.4	0.6	18.9

*Among the types of activities included in "other" there are pharmaceutical, chemical, oil and gas industries, and other industries not previously specified.

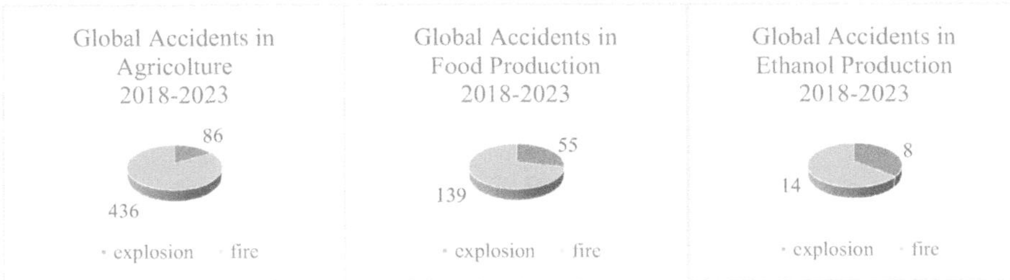

FIGURE 2.12 Number of global accidents for a specific kind of activity [24–29].

Concluding, it is possible to say that the statistical analysis here reported emphasizes how each individual piece of data should be considered a mere reference and not a truly representative value.

The general impossibility of providing a detailed and constantly updated snapshot of the actual scenario through the use of statistics stems from several intrinsic elements of the statistical tool, such as: (a) industrial evolution (during the time period under analysis, it is possible to observe the opening and closing of various business activities and their transfers. It is, therefore, possible that less detailed statistics may suffer from the presence of partially representative trends [30]); (b) non comparable definitions (different sources may use different definitions or inclusion parameters, both in reference to the events and the starting conditions, generating non-overlapping statistical values); (c) regulatory differences (if the statistics include data from different countries, it is possible that the reference regulations are structured in a non-homogeneous way regarding the obligations and minimum information to be reported in the event of an event useful for the analysis [31]); (d) under reports (many accidents, especially the more limited ones, do not appear in official sources. The available figures, therefore, tend to underestimate the actual frequency [31, 32]); (e) changes in reporting practices and surveillance over time (a progressive improvement of practices and tools used for reporting activity leads to a potential increase in the number of events over time for greater capacity for reporting and recording events [33]).

It is worth noticing that most data reported in Section 2.3 regard the food industry (or industries with a high presence of food powders), rather than the pharmaceutical one. The reason was dictated by a series of objective conditions, especially in relation to what is highlighted in point (b) of the previous list. Some accidents that occurred in chemical plants producing raw materials/excipients for pharmaceutical industries have been classified as "chemical industrial accidents," but, for the purposes of the present text, they should be considered as pharmaceutical industrial accidents. Lacking, at present, further detailed information useful for developing an appropriate statistical analysis for the pharmaceutical industry, it was deemed useful to focus mainly on food industry (or industries involved in the handling of food powders).

TABLE 2.21
Combustible Involved in Dust Fire or Dust Explosion 2016–2023 [24–29]

YR	Food Prod (%)	*F*	*E*	Not known (%)	*F*	*E*	Wood Prod (%)	*F*	*E*	Metal (%)	*F*	*E*	Other* (%)	F	E
2016	33.0	/	/	3.0	/	/	30.0	/	/	9.0	/	/	24.0	/	/
2017	32.0	/	/	8.0	/	/	30.0	/	/	10.0	/	/	20.0	/	/
2018	43.0	/	/	5.0	/	/	28.0	/	/	10.0	/	/	12.0	/	/
2019	40.0	107	24	9.0	25	5	26.0	62	24	11.0	29	6	13.0	27	16
2020	48.0	80	28	7.1	14	2	27.6	47	15	7.6	10	7	9.8	14	8
2021	49.5	91	16	4.6	5	5	25.0	35	14	6.0	10	3	14.8	22	15
2022	48.4	84	19	6.6	10	4	25.4	43	9	9.4	13	7	10.3	13	11
2023	43.7	115	23	4.7	11	4	35.4	93	12	5.7	13	5	10.4	31	9

*Among the types of fuels included in "others," there are coal, paper, plastic, carbon, sulfur, and other not previously specified.

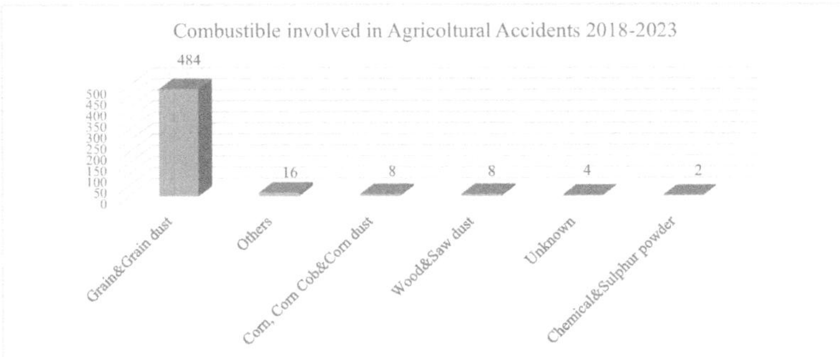

FIGURE 2.13 Combustible involved in agricultural accidents 2018–2023 [24–29].

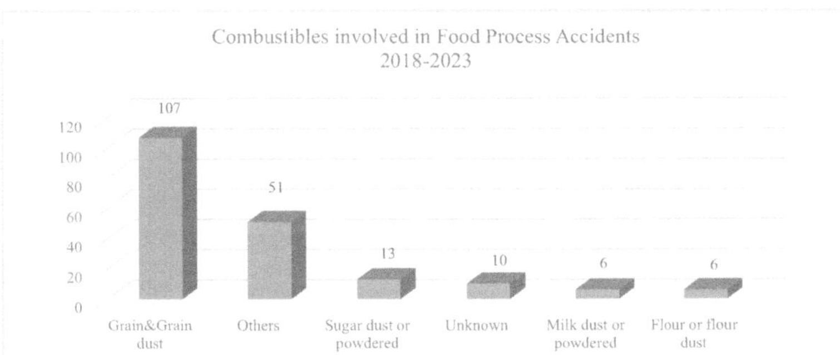

FIGURE 2.14 Combustibles involved in food process accidents 2018–2023 [24–29].

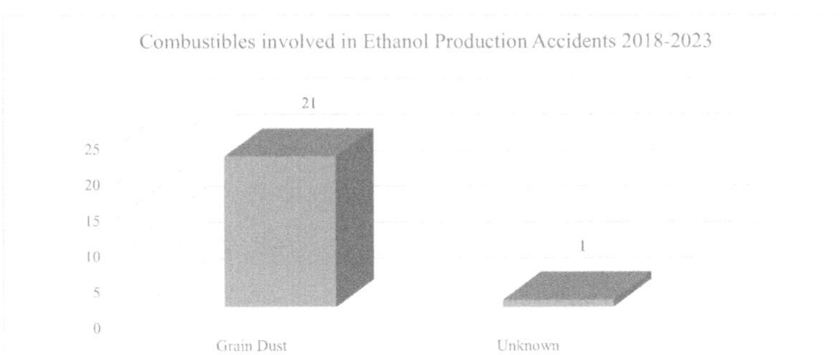

FIGURE 2.15 Combustibles involved in ethanol production accidents 2018–2023 [24–29].

TABLE 2.22
Combustibles Cited as "Others" in Figures 2.13 and 2.14 [24–29]

Sector	Combustible	Number of Events
Agricultural	Straw dust	1
	Haylage	1
	Hay dust	1
	Fertilizer	2
	Sweet potato	1
	Cotton dust	3
	Sunflower	1
	Coal dust	1
	Silage	2
	Flax dust	1
	Bone metal	1
	Seed dust	1
Food Process	Pecan	1
	Spices	2
	Cocoa dust	3
	Cotton seed	2
	Cereal dust	1
	Magnesium dust	1
	Protein dust	1
	Sunflower	1
	Coal dust	1
	Barley/malt dust	4
	Wood or saw dust	5
	Fishmeal dust	3
	Canola meat dust	1
	Rice dust or dried rice	3
	Fruit seed dust	1
	Blood meal	1
	Baby formula	1
	Powdered chocolate	1
	Palm kernel dust	1
	Peanut, nut, or almond dust	5
	Dried egg dust	1
	Chile dust	1
	Overdried pulp	1
	Soybean dust	1
	Potato dust	1
	Animal food	2
	Dehydrated Alfalfa	2
	Dried herbs	1
	Coffee dust	1
	Pomace dust	1

2.4 ECONOMIC LOSSES AND ASSOCIATED COSTS

Dust explosions within industrial settings have wide-ranging repercussions that extend beyond safety and operational concerns, significantly impacting the economic sector. These catastrophic events can generate economic losses generally categorized into two main groups: direct economic damages and indirect economic damages.

Direct damages are typically associated with the immediate and tangible financial outlay needed to repair or rebuild physical infrastructure damaged by the explosion. Indirect damages, on the other hand, are more complex to quantify and involve losses such as production downtime, reputational harm among stakeholders, and long-term financial instability [34]. While direct costs are more easily traceable to a single accident, such as an explosion, indirect costs are less visible and harder to estimate. This distinction is crucial when evaluating the true financial burden of an industrial disaster, as indirect consequences often outweigh direct ones. Therefore, implementing appropriate cost estimation tools to assess the potential impact of such events becomes vital for corporate governance. This not only promotes awareness of process and occupational safety investment but also enhances resilience to future disruptions [35].

Companies have both economic and ethical incentives to invest in safety systems aimed at reducing the likelihood and severity of catastrophic events. Dust explosion-related costs vary depending on the context and scale of the accident. While some expenses are immediately evident, others are more subtle yet significantly influence the total economic impact (in extreme cases, particularly those involving fatalities and environmental liability, single dust explosions have been estimated to result in damages of over $1 million [36]). Particularly, the most relevant cost categories associated with dust explosions are:

(A) Legal Liabilities
Legal proceedings often represent one of the most significant financial burdens following a dust explosion. Depending on the severity and consequences of the event, lawsuits can be civil, criminal, or both (potentially amounting to several million dollars/euros). In severe cases, legal liabilities alone have led to corporate bankruptcy.

(B) Lost Revenue from Downtime
Lost income due to halted operations can be estimated using a continuous production model. For example, a company generating $50 million annually that ceases operations for just three days could suffer losses of approximately $430,000. This estimate assumes a relatively limited incident with quick resumption of activity.

(C) Reconstruction Costs
Rebuilding costs range from tens of thousands to several million dollars. According to data from dustsafetyscience.com [36], 69% of reported incidents resulted in reconstruction expenses exceeding $100,000 per explosion.

(D) Worker Compensation
Medical and compensatory costs for injured workers can also be substantial. In severe cases involving hospitalization, expenses may exceed $1 million. Even in less severe scenarios, costs typically start around $200,000.

(E) Fines and Penalties
Fines depend on the nature and extent of regulatory violations. Intentional breaches that contributed to the explosion carry far higher penalties (often exceeding $1 million) compared to non-intentional but significant infractions, which may still incur penalties in the tens of thousands of dollars [36].

(F) Insurance Premium Increases
Post-incident insurance premiums often rise by 10% to 30%, meaning that a $1 million annual policy could see an increase of $100,000 to $300,000 [36].

Beyond company-specific losses, externalized costs often affect the surrounding community, contributing to broader social and environmental impacts. These can include air and soil contamination from hazardous substance release, requiring extensive and costly remediation efforts.

One notable study by Guignet et al. (2023) [37], published by the US Environmental Protection Agency (EPA), investigated the long-term economic fallout of chemical industrial accidents on neighboring communities, focusing specifically on property devaluation resulting from explosions, fires, and hazardous releases. By analyzing 661 industrial incidents across the United States between 2004 and 2019, and comparing them to millions of property transactions, the study revealed that 12% to 55% of the US population (i.e., between 40 and 177 million people) potentially live within zones at risk from industrial disasters. These findings also highlighted persistent social and environmental inequalities, with marginalized communities disproportionately affected by proximity to hazardous industrial sites. The study also found that industrial accidents with off-site consequences (evacuations, structural damage, injuries) caused a 2–3% reduction in residential property values within a 5.75 km radius. Even more critically, this devaluation persisted for over 10–15 years, negatively impacting local economies and the financial stability of affected families. Average property losses exceeded $5,000 per home, and cumulative property value loss across all 661 incidents was estimated at $39.5 billion. Interestingly, the majority of this economic burden was caused by just 25% of the events, reinforcing the notion that a small number of major incidents drive most long-term economic damage. Events confined within industrial perimeters, with no significant off-site consequences, had negligible effects on local property markets [37]. The findings of Giugnet et al. go beyond the chemical sector and are equally applicable to dust explosion scenarios. Investments in plant modernization, worker training, and effective risk management may seem financially burdensome at first, but when weighed against the potentially avoidable social and economic costs, they should be reframed as community-wide and environmental investments.

Ultimately, industrial disasters such as dust explosions have a profound and lasting economic footprint, both for the affected company and for the broader community. It is essential to recognize that safety expenditures must be viewed as strategic investments, capable of generating a positive return on investment, sometimes even resulting in long-term profitability for the organization.

REFERENCES

1. Kletz T., 2001. *Learning from Accidents*. 3rd ed. Gulf Professional Publishing, Oxford.
2. Chemical Safety Board (CSB) report 2003-07-I-NC SEPTEMBER 2004, last accessed on 01/15/2026.
3. Rao B.K., Rao G.V., Reddy V.K., 2021. Critique of final enquiry report on fire and explosion at VNAI, Chandrapadia, last accessed on 01/15/2026.
4. Kashyap K., 2024. The rise of India's pharmaceutical industry to a forecasted $450 billion. Forbes.com, November 2024. https://www.forbes.com/sites/krnkashyap/2024/11/16/the-rise-of-indias-pharmaceutical-industry-to-a-forecasted-450-billion/, last accessed 01/15/2026.
5. Safety data Sheet (SDS) Methanol, Roth. Date of compilation: 2016-04-29 Revision: 2024-09-18, last accessed on 01/15/2026.
6. Safety data Sheet (SDS) Apronal, Cayman Chemical. Printing date 10/26/2023 Revision date 10/26/2023.
7. Cloney C., 2019. Explosion at pharmaceutical manufacturing plant. dustsafetyscience.com, August 2019. https://dustsafetyscience.com/dust-explosion-andhra-pradesh-india/, last accessed 01/15/2026.
8. Cloney C., 2020. Five killed and one injured in explosion at pharmaceutical plant. dustsafetyscience.com, September 2020. https://dustsafetyscience.com/potential-dust-explosion-tianmencity-china/, last accessed 01/15/2026.
9. Cloney C., 2021. Five workers injured in explosion at pharmaceutical manufacturer. dustsafetyscience.com, October 2021. https://dustsafetyscience.com/tablet-dryer-explosion-baddi-india/, last accessed 01/15/2026.
10. Viswanath P., 2025. Sigachi blast exposes recurring safety failures and worker deaths in Indian factories. Business and Human Right Centre, July 2025. https://www.business-humanrights.org/en/latest-news/india-families-allege-recurring-safety-failures-after-sigachi-industries-plant-blast-kills-migrant-workers-investigation-ongoing/, last accessed 01/15/2026.
11. ARIA report n. 8781 (2009), last accessed on 01/15/2026.
12. Chemical Safety Board (CSB) report 2008-05-I-GA SEPTEMBER 2009, last accessed on 01/15/2026.
13. INSEE, 2025. Transformation coefficient from one year's Euro or French Franc to another year's Euro or French Franc – Base 2015 – Stopped series. https://inflation-history.com/it-IT/?currency=FRF&amount=70000000&year=1982, last accessed 01/15/2026.
14. Jie Y., 2010. Dust blast at workshop kills 19. China Daily, February 2010. https://www.chinadaily.com.cn/china/2010-02/26/content_9506519.htm, last accessed 01/15/2026.
15. Cloney C., 2019. Three people injured in dust explosion at Dutch spice factory. dustsafetyscience.com, September 2011. https://dustsafetyscience.com/dust-explosion-wijhe-netherlands/, last accessed 01/15/2026.
16. Raj S., 2022. Andhra Pradesh: Two dead, 9 injured in sugar factory explosion in Kakinada. Times of India, August 2022. https://timesofindia.indiatimes.com/city/visakhapatnam/andhra-pradesh-two-dead-9-injured-in-sugar-factory-explosion-in-kakinada/articleshow/93672098.cms, last accessed 01/15/2026.
17. ATEX Explosion Hazards. Dust explosion at Turkish pasta factory kills one, injures 30. February 2025. https://www.explosionhazards.co.uk/dust-explosion-at-turkish-pasta-factory-kills-one-injures-30/, last accessed 01/15/2026.

18. Taddia M., 2005. Il conte Morozzo e la rivoluzione chimica. *La Chimica e L'industria*, 2(87), 66–68.
19. Tosti T., 2021. Esplosioni di polveri di lana nell'industria tessile. Master's degree thesis. Chemical and Sustainable Process Engineering. Polytechnic of Turin.
20. Yuan Z., Khakzad N., Khan F., Amyotte P., 2015. Dust explosions: A threat to the process industries. *Process Safety and Environmental Protection*, 98, 57–71.
21. Amyotte P., Khan F.I., 2019. *Methods in Chemical Process Safety – Dust Explosions*. 1st ed. Academic Press, Elsevier, Cambridge.
22. Chemical Safety Board (CSB) report 2006-H-1 November 2006, last accessed on 01/15/2026.
23. Yan Q.X., Yu J.L., 2012. Dust explosion incidents in China. *Process Safety Progress*, 31. DOI: 10.1002/prs.11482.
24. Cloney C., 2019. 2018 Combustible dust incident report – Version #1. DustEx Research Ltd. Retrieved from http://dustsafetyscience.com/2018-Report, last accessed on 01/15/2026.
25. Cloney C., 2020. 2019 Combustible dust incident report – Version #1. DustEx Research Ltd. Retrieved from http://dustsafetyscience.com/2019-Report, last accessed on 01/15/2026.
26. Cloney C., 2021. 2020 Combustible dust incident report – Version #1. DustEx Research Ltd. Retrieved from http://dustsafetyscience.com/2020-Report, last accessed on 01/15/2026.
27. Cloney C., 2022. 2021 Combustible dust incident report – Version #1. DustEx Research Ltd. Retrieved from http://dustsafetyscience.com/2021-Report, last accessed on 01/15/2026.
28. Cloney C., 2023. 2022 Combustible dust incident report – Version #1. DustEx Research Ltd. Retrieved from http://dustsafetyscience.com/2022-Report, last accessed on 01/15/2026.
29. Cloney C., 2024. 2023 Combustible dust incident report – Version #1. DustEx Research Ltd. Retrieved from http://dustsafetyscience.com/2023-Report, last accessed on 01/15/2026.
30. OECD, 2007. Eurostat – OECD manual on business demography statistics. European Communities / OECD.
31. International Labour Office, 2012. Improvement of national reporting, data collection and analysis of occupational accidents and diseases. International Labour Office, Geneva.
32. Nifuku M., Matsuda T., Enomoto H., 2000. Recent development of standardization of testing methods for dust explosion in Japan. *Journal of Loss Prevention in the Process Industries*, 13, 243–251.
33. EUROSTAT, 2021. European Statistical System handbook for quality and metadata reports 2021 re-edition. Publications Office of the European Union, 2021. DOI: 10.2785/616374 KS-GQ-21-021-EN-N.
34. Santana D.A.J., Arana C.Y., Gòmez G.O., Furka D., Furka S., Orozco L.J., Di Benedetto A., Russo D., Portarapillo M., Febles S.J., 2022. Fire and Explosion Economic Losses (FEEL) index: A new approach for quantifying economic damages due to accidents in hydrocarbon storage sites. *Process Safety and Environmental Protection*, 165, 77–92.
35. Gavious A., Mizrahi S., Shani Y., Minchuk Y., 2009. The costs of industrial accidents for the organization: Developing methods and tools for evaluation and cost–benefit analysis of investment in safety. *Journal of Loss Prevention in the Process Industries*, 22, 434–438.
36. Cloney C., 2019. What is the potential cost of a dust explosion incident with Chris Giusto. dustsafetyscience.com, September 2019. https://dustsafetyscience.com/dust-explosion-cost-chris-giusto/, last accessed 01/15/2026.
37. Guignet D., Jenkins R.R., Nolte C., Belke J., 2023. The external costs of industrial chemical accidents: A nationwide property value study. NCEE Working Paper 23-01.

3 Dust Explosion Phenomenon

3.1 CHEMICAL AND PHYSICAL BASIS OF DUST EXPLOSIONS

Dust explosions represent a significant hazard in various industrial settings where fine particulate matter is handled, processed, or stored. These events occur when a combustible dust is dispersed in air in sufficient concentration and encounters an ignition source, leading to rapid combustion and a sudden release of energy. Unlike gas or vapor explosions, dust explosions present unique challenges due to the heterogeneous nature of the fuel, the influence of particle size and shape, and the dynamics of dispersion within confined or semi-confined environments.

Industries such as pharmaceuticals, food processing, woodworking, agriculture, and chemical manufacturing are particularly vulnerable to dust explosion hazards. Historical accidents have highlighted the potentially devastating consequences of such explosions (see Chapter 2), including loss of life, extensive property damage, and prolonged operational disruptions. Despite increased awareness and regulatory measures, dust explosions continue to occur, emphasizing the need for a thorough understanding of the underlying mechanisms.

At the heart of a dust explosion lies a complex interplay of physical and chemical phenomena. Factors such as dust concentration, particle characteristics, oxygen availability, ignition energy, and confinement conditions all contribute to the likelihood and severity of an explosion. The phenomenon typically unfolds in two stages: a primary explosion at the site of ignition and, in many cases, a more powerful secondary explosion triggered by the dispersion of settled dust.

The following sections delve into the fundamental physical and chemical principles governing dust explosions. Particularly, they aim at providing a comprehensive framework for understanding how and why these events occur, forming the basis for effective prevention and mitigation strategies.

3.1.1 Generalities on Explosion Phenomenon

An explosion is defined as a release of energy within a sufficiently short time and in a sufficiently small volume to generate a pressure wave of finite magnitude that propagates away from the point of release and can be heard.

DOI: 10.1201/9781003645320-3

The basic phenomenology of an explosion can be easily described by referring to the decomposition of a solid in an open or confined environment, as illustrated in Figure 3.1.

Consider an unstable solid to which enough energy is supplied to initiate a decomposition reaction. Decomposition reactions, like combustion processes, are highly exothermic and involve the rapid conversion of the solid material into high-temperature gaseous decomposition products. As a result, within a practically instantaneous timescale, a given quantity of matter that initially occupied a volume V, characteristic of a high-density solid, finds itself transformed into a gas phase whose potential volume of occupation is approximately 1000 times greater, because gases typically exhibit a density around 1/1000 that of their solid counterparts. The decomposition gases, therefore, undergo violent expansion, generating pressure waves that propagate through the surrounding atmosphere.

It is important to note that the generation of decomposition gases is not strictly instantaneous; rather, the gas flow rate varies over time, increasing progressively until it reaches a peak. This trend is attributed to the temperature rise in the reaction zone as the process evolves. Consequently, a sequence of pressure waves, or gas expansion fronts, is formed, propagating through the medium at a velocity v corresponding to the local speed of sound under the prevailing temperature conditions ($c(T)$).

This implies that, since temperatures tend to rise to a maximum during the decomposition process, the pressure waves formed in the moments immediately following ignition (that is, at temperature equal to T_0) will tend to propagate at progressively higher velocities (see Figure 3.1), eventually collapsing into a single, well-defined front known as a shock wave front. This front travels at the speed of sound corresponding to the maximum temperature attained during the decomposition reaction, denoted as $c(T_{max})$.

Such a phenomenon can also develop in the context of combustion reactions or reactions involving both decomposition and combustion, and it may involve solids

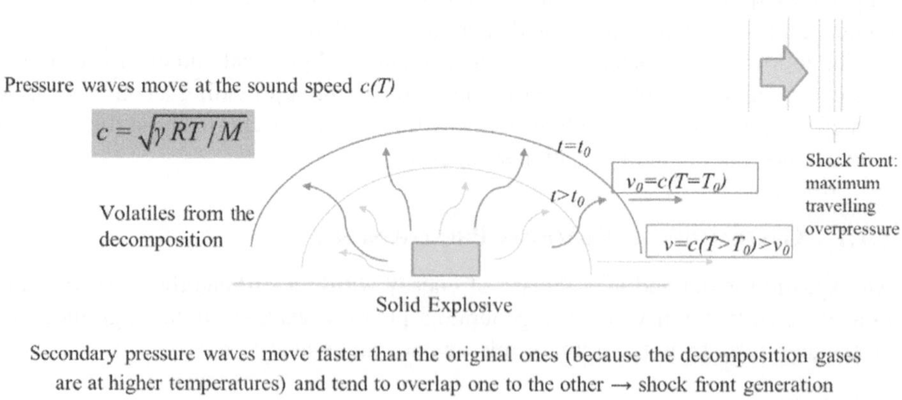

FIGURE 3.1 Sketch of the dynamics of an explosion derived by a solid decomposition.

(e.g., explosives or combustible dusts), liquids (typically in the form of mists), or flammable gases and vapors. Since industrial explosions of concern typically occur in the presence of atmospheric air, the following definition, referring to *organic combustible dusts*, is commonly adopted for an explosive atmosphere.

An explosive atmosphere is a mixture with air, under atmospheric conditions, of flammable substances in the form of gases, vapors, mists, or dusts, which can ignite and propagate combustion when an ignition source is introduced.

Within this context, an explosion occurs when the composition of the mixture meets at least one of the following criteria:

a) Minimum ignition temperature is reached, that is, the minimum temperature at which a solid (even in the form of fine particles) releases sufficient gases to form an ignitable mixture with air in the presence of an ignition source (organic combustible dusts).
b) Explosion Limits are reached. Explosion limits define the concentration range within which a mixture (dust and air) is capable of exploding. Two limits are identified: the Lower Explosion Limit (LEL) and the Upper Explosion Limit (UEL).

The Lower Explosion Limit (LEL) is the minimum "fuel" concentration in air below which an explosion cannot occur. Below this threshold, particles (e.g., dust) are too sparsely distributed for the heat released by the oxidation of individual particles to ignite neighboring ones, thereby preventing the propagation of a self-sustaining reaction.

The Upper Explosion Limit (UEL) is the maximum "fuel" concentration in air above which an explosion cannot occur. Beyond this concentration, the oxidizing agent (e.g., oxygen) becomes insufficient to support the release of enough energy to propagate the reaction.

The minimum and maximum explosible concentrations depend on several variables, including surface condition of solid particles, particle size, temperature, pressure, presence of inert gases (e.g., N_2, CO_2) or inert dusts, and oxygen concentration.

It is important to emphasize that confinement significantly exacerbates the effects of an explosion, as the resulting overpressures are considerably greater than those observed in unconfined environments.

3.1.2 Classification of Explosions

The task of classifying any given phenomenon is inherently complex and somewhat artificial. However, for the purpose of this book, it is considered useful to categorize the various types of explosions that may occur in the context of a major accident into two broad families: physical explosions and chemical explosions.

Physical explosions occur because of the sudden rupture of a vessel containing a substance under elevated pressure. The conditions leading to such internal overpressure within the vessel are typically identified through hazard assessment and incident scenario analysis procedures (refer to Chapter 5).

The mechanism underlying the formation of a shock wave in a physical explosion can be attributed to the difference between the internal energy of the substance within the vessel prior to rupture and its energy state in the surrounding environment post-rupture. This energy difference can be estimated by assuming that the substance undergoes a violent isentropic expansion during the collapse of the equipment.

In this context, the energy released, q, can be approximated by the following equation (3.1):

$$q \approx \frac{(P_0 - P_a) \cdot V_0}{\gamma - 1} \tag{3.1}$$

where V_0 [m³] is the internal volume of the vessel, P_0 [Pa] is the internal pressure, P_a [Pa] is the external (atmospheric) pressure, and γ [–] is the isentropic expansion coefficient of the substance.

Under real accident conditions, the portion of internal energy effectively converted into a shock wave typically ranges from 40% to 80% of the maximum energy available, as calculated using (3.1).

Chemical explosions, on the other hand, result from the ignition of exothermic chemical reactions, such as combustion or decomposition of reactive substances. A combustion or decomposition reaction becomes explosive when heat exchange with the environment is nearly negligible (i.e., under pseudo-adiabatic conditions). Under such circumstances, the entirety of the thermal energy released by the reaction is used to raise the temperature of the reaction zone, which, in turn, accelerates the reaction rate. This leads to an exponential increase in system temperature, which can only be halted upon depletion of the reactants.

Chemical explosions can be classified into two distinct phenomena based on the propagation speed of the flame front (in the case of combustion reactions) through the medium (typically air):

- *Deflagration.* A deflagration is an explosion in which the reaction propagates via thermal conduction and the diffusion of unreacted substances toward the reaction front. The flame front travels at subsonic velocity, and the overpressures generated across the shock front are relatively modest.
- *Detonation.* A detonation is an explosion in which the reaction propagates through the medium via shock wave-driven ignition of the reactive mixture, sustained by the autoignition temperature. In this case, the pressure waves themselves initiate the reaction, creating a supersonic flame front that moves along with the shock wave. The resulting overpressures across the shock front are significantly higher.

While deflagrations are relatively common for gas and dust mixtures, even in unconfined environments, detonations generally require either a high degree of confinement or, in open-air scenarios, the release of large quantities of combustible material.

3.1.3 Dust Explosions Generalities

A dust explosion is an exothermic phenomenon that occurs because of a violent combustion reaction between a combustible dust and atmospheric air. For such an explosion to occur, the dust must be suspended in air, forming a dust cloud.

Much like fires, which follow the well-known "fire triangle" (fuel, oxidizer, and ignition source), dust explosions are commonly described referring to the so-called "dust explosion pentagon," which outlines the conditions required for the development of an explosive event within a dust cloud (see Figure 3.2). Particularly, the following five conditions must be simultaneously satisfied for a dust explosion to occur:

- Presence of a combustible dust, such as: (a) *natural organic substances*: grains, sugar, coal, etc.; (b) *synthetic organic materials*: pesticides, plastics, etc.; (c) *oxidizable metals*: aluminum, zinc, iron, magnesium, etc.
- Sufficient oxygen concentration (oxidizer) in the atmosphere to sustain the combustion/oxidation reaction.
- Presence of an ignition source, which may include: (a) open flames (capable of providing energy exceeding the minimum ignition energy, thereby initiating oxidation within the dust cloud); (b) welding and cutting operations (which generate incandescent particles that can travel significant distances and ignite dust clouds); (c) friction or impact (mechanical contact between surfaces may cause heat buildup and spark generation); (d) hot surfaces (if the surface temperature exceeds the autoignition temperature of the dust, ignition of either clouds or dust layers may occur); (e) electrical and electrostatic sparks (caused by equipment malfunction or circuit operation, these sparks often possess sufficient energy to trigger a dust cloud explosion).

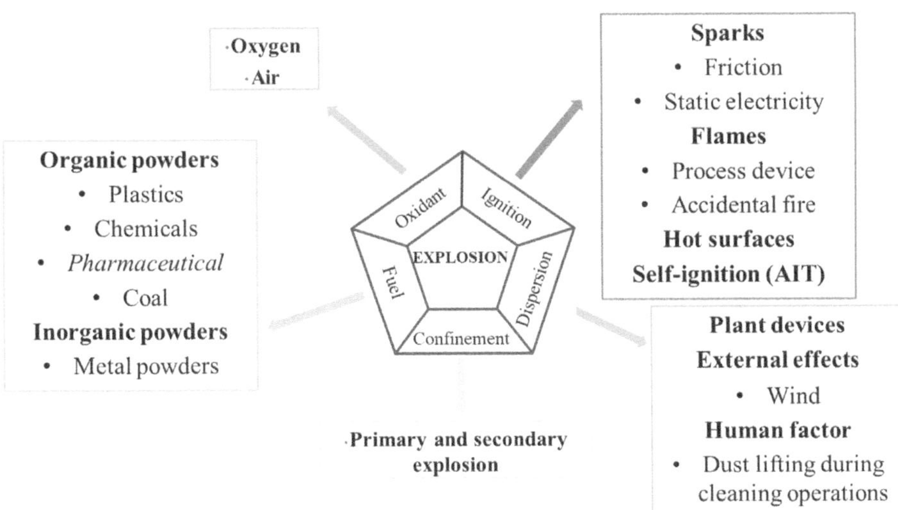

FIGURE 3.2 Dust explosion pentagon.

- Confinement of the dust cloud, for example, within a room or enclosed process equipment.
- Adequate mixing of the fuel (dust) and oxidizer (air), typically achieved by dispersion that forms a cloud of fine particles.

If any one of these five conditions is absent, a dust explosion cannot take place.

3.1.4 Key Factors Influencing Dust Explosivity

The explosibility of dust clouds is determined by a range of interdependent physical, chemical, and environmental factors. These factors influence not only the likelihood of ignition but also the rate of flame propagation, the peak explosion pressure, and the extent of damage caused by the explosion. A comprehensive understanding of these factors is essential for evaluating explosion risks and designing effective mitigation strategies. The primary factors are detailed in the following.

3.1.4.1 Chemical Nature of the Dust (Fuel Composition)

Each combustible dust has a distinct chemical structure, elemental composition, and heat of combustion, all of which strongly influence its reactivity. Organic dusts (such as starch, flour, sugar, cellulose) tend to exhibit moderate explosion violence, whereas metallic dusts (e.g., aluminum, magnesium, titanium) often produce high-temperature reactions with elevated flame speeds and overpressures, due to their high enthalpies of oxidation. Furthermore, some dusts may undergo complex decomposition reactions that generate additional reactive intermediates, further enhancing explosion severity.

3.1.4.2 Particle Size and Specific Surface Area

The granulometry of the dust is a critical determinant of explosibility. Smaller particles have a higher surface-area-to-volume ratio, which facilitates more efficient heat and mass transfer between the dust and the surrounding oxidizer. This leads to faster ignition, higher reaction rates, and shorter induction periods. In contrast, coarser particles ignite less readily and require higher dust concentrations to sustain combustion. Particle morphology (e.g., spherical vs fibrous) and the presence of fines also affect dispersion characteristics and flame propagation.

3.1.4.3 Surface Condition of Particles

Dust particles that are oxidized, passivated, or coated with protective films tend to be less reactive. For example, metal dusts exposed to air often form oxide layers that inhibit rapid oxidation. Such surface passivation increases the minimum explosible concentration (MEC) and delays ignition. On the other hand, freshly ground or atomized particles with clean, reactive surfaces exhibit much higher explosion sensitivity.

3.1.4.4 Oxygen Concentration in the Atmosphere

The oxidizer concentration is vital to sustaining a combustion reaction. In air (typically 21% oxygen), many dusts are readily ignitable. However, increasing the oxygen level (e.g., through enrichment) leads to faster flame speeds, higher pressure rise rates, and lower ignition thresholds. Conversely, in oxygen-deficient environments, combustion is progressively suppressed. For most organic dusts, the critical oxygen concentration below which explosions no longer occur lies between 7% and 11%.

3.1.4.5 Temperature of the System

Elevated ambient or pre-ignition temperatures accelerate chemical kinetics, reduce ignition delays, and bring the system closer to the autoignition temperature of the dust-air mixture. High temperatures reduce the thermal inertia required to initiate and propagate the flame front, thereby expanding the explosibility range. In practice, even modest temperature increases can significantly reduce the minimum ignition energy (MIE) and facilitate flame spread.

3.1.4.6 Initial Pressure

Higher initial system pressures lead to more violent explosions. The pressure not only increases the concentration of reactants but also amplifies the peak explosion pressure, as the final pressure is superimposed on the pre-existing pressure. Additionally, high-pressure environments promote faster heat transfer, greater reaction rates, and shorter flame acceleration distances. Therefore, confined and pressurized vessels present a particularly high explosion risk.

3.1.4.7 Humidity and Moisture Content

Moisture in the dust or surrounding atmosphere generally has an inhibiting effect on explosibility. Water promotes agglomeration of particles, reducing their dispersibility, and effective surface area. Moreover, during combustion, water undergoes phase transition (evaporation), which absorbs latent heat, thereby cooling the reaction zone and hindering ignition. However, excessive moisture may also lead to uneven drying, creating local hot spots that reintroduce risk.

3.1.4.8 Turbulence and Mixing Intensity

Turbulence greatly influences combustion dynamics by enhancing the mixing of fuel and oxidizer. A highly turbulent dust cloud will exhibit a more wrinkled and unstable flame front, leading to faster propagation, shorter deflagration-to-detonation transition distances, and greater overpressures. Sources of turbulence include fans, pneumatic transport systems, or sudden mechanical disturbances.

3.1.4.9 Presence of Inert Gases (e.g., N_2, CO_2)

Inert gases dilute the reactive mixture and "absorb" thermal energy, thereby increasing the ignition energy requirement and reducing the peak explosion pressure. They tend to lower the upper explosion limit (UEL) while having little to no effect on the Lower Explosion Limit, thus narrowing the explosibility range. CO_2 also plays a chemical inhibition role by interfering with radical chain reactions in the flame zone (ref to organic dusts).

3.1.4.10 Presence of Inert Dusts

Inert particulate matter, such as silica, calcium carbonate, or diatomaceous earth, does not participate in combustion reactions but still impacts explosion behavior. When mixed with combustible dusts, inert particles increase the MEC and reduce flame speed by diluting the fuel load and absorbing heat. Their addition is a common mitigation strategy in industrial settings (e.g., rock dusting in coal mines).

3.1.4.11 Hybrid Mixtures (Dust and Flammable Gases or Vapors)

The presence of flammable gases or vapors in combination with combustible dusts creates hybrid mixtures, which are typically more hazardous than either component alone. These mixtures often exhibit (a) lower minimum ignition energies (MIE), (b) reduced autoignition temperatures, (c) expanded flammability limits, (d) increased maximum pressure and rate of pressure rise.

Hybrid mixtures are particularly dangerous because even a sub-threshold dust concentration may become explosible when accompanied by a flammable vapor and vice versa.

The comprehensive understanding of the role of explosibility parameters (see Chapter 4) is essential for the assessment of dust explosion hazards, selection of safe operating conditions, and the design of effective prevention and protection systems in compliance with international standards (see Chapter 5).

3.2 DYNAMICS OF DUST EXPLOSIONS

Combustible dust explosions can be controlled by two different combustion mechanisms: homogeneous and heterogeneous. In general, the combustion of carbonaceous dusts and low-temperature vaporizing particles belongs to a homogeneous mechanism, while metal dusts burn through heterogeneous surface oxidation [1–4]. As reported in Figure 3.3, homogeneous combustion mechanism usually involves steps such as: (a) particles heating and water vaporization, (b) devolatilization of unburnt compounds, (c) volatile gases–air mixing and, (d) homogeneous combustion [5]; while heterogeneous combustion proceeds mainly through: (a) oxygen diffusion into either a molten or a solid core phase (into the dust particle) and (b) heterogeneous oxidation.

The generation of volatile gases is a process characteristic of the homogeneous mechanism and, therefore, it can be considered as a criterion to distinguish between homogeneous and heterogeneous combustion. Accordingly, thermogravimetric analysis (TGA) can be used to discriminate the combustion mechanisms of different dust samples [4]. For a given test, a small dust sample has to be placed into a TGA furnace and heated at a given heating rate (e.g., 5°C/min, from room temperature to 800°C) under both nitrogen and air atmosphere. The mass of the sample is recorded during the experiment, as shown in the example of Figure 3.4 for a carbonaceous dust (sweet potato) under both nitrogen and air atmosphere.

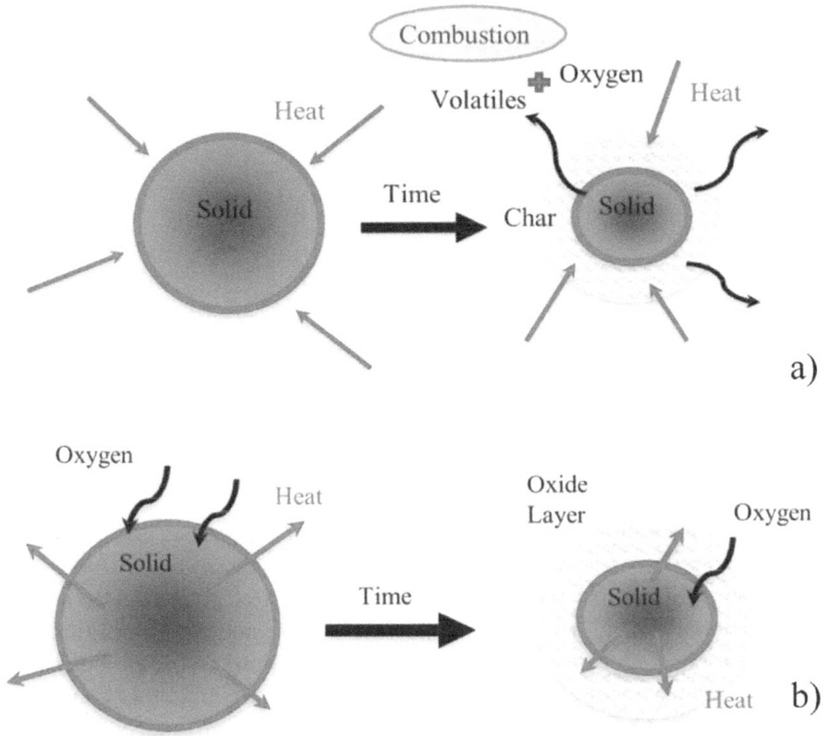

FIGURE 3.3 Schematic representation of both homogeneous (a) and heterogeneous (b) combustion processes.

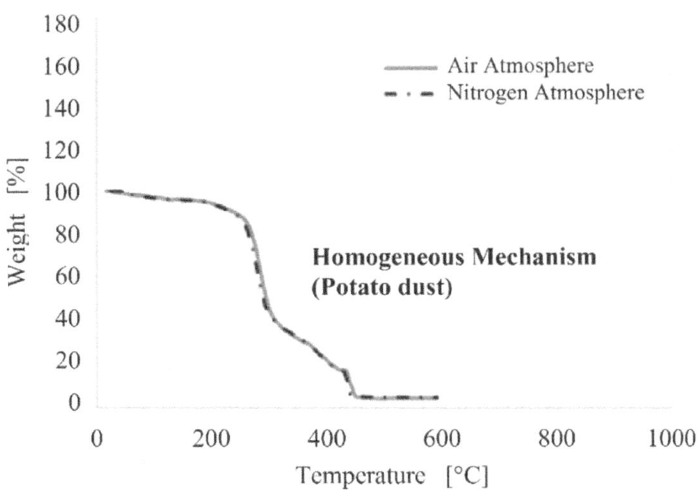

FIGURE 3.4 TGA of a carbonaceous dust (sweet potato) under both nitrogen and air atmosphere.

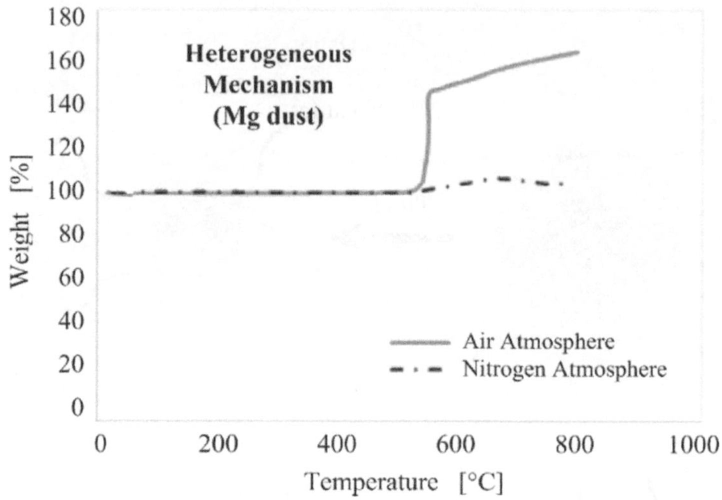

FIGURE 3.5 TGA for Mg dust under both nitrogen and air atmosphere.

It is possible to identify four successive steps: (a) loss of physically adsorbed water, up to about 240°C, (b) starch depolymerization and decomposition, up to about 320°C, (c) decomposition of hemicelluloses, cellulose and some lignin, up to about 430°C, and (d) decomposition of the remaining lignin and other components, up to about 442°C. The similarity of the TGA profiles in air and N_2 atmosphere shows that the weight loss of sweet potato is nearly independent of the oxygen presence. Therefore, the deflagration mechanism of sweet potato dusts will be governed by a homogeneous combustion, occurring through a sequential process starting with the dust devolatilization followed by a vapor-phase combustion [6, 7].

On the contrary, Figure 3.5 also gives an example of a TGA test for magnesium dusts under both nitrogen and air atmosphere.

As it can be seen, under nitrogen atmosphere, the TGA trace shows, at first, a small mass increase followed by a little decrease mainly due to the formation of Mg_3N_2. However, under air atmosphere, a consistent weight increment is clearly identified, which corresponds to the surface heterogeneous oxidation. The oxide coating is produced via the formation of a Mg-O mixture followed by an exothermic phase separation within the burning particle. In this case, the deflagration mechanism of magnesium particles will be governed by the surface heterogeneous oxidation [8, 9].

Moreover, there is another difference between metallic and organic dust explosions [10]. Metallic dust clouds can be ignited by the radiation coming from another dust cloud where burning is in progress [11], while organic powders cannot. This is because the intrinsic inability of the organic dust clouds to self-sustain the propagation of a flame in air at normal temperature and pressure, in contrast to metallic powders.

3.2.1 Homogeneous Mechanism

As already stated, homogeneous combustion involves the following main steps:

(1) Particle Heating and Moisture Evaporation
The process begins with the heating of suspended dust particles when they are exposed to an external ignition source (e.g., flame, spark, hot surface). Heat transfer occurs primarily via convection from the hot gases (external heat transfer) and radiation from the ignition source or other reacting particles, then, through conductive heat fluxes into the particle (internal heat transfer).
Water evaporation is the first mass transfer process to occur if the particles contain physically adsorbed moisture or bound water. The latent heat of vaporization of water ($\approx$ 2.26 MJ/kg) must be overcome, which absorbs thermal energy and temporarily slows down the temperature rise of the particle. This step is endothermic, and the presence of water acts as a thermal buffer, reducing the risk of immediate ignition. However, once water is removed, the temperature rise accelerates.

(2) Devolatilization of Organic Components
After dehydration, the particle temperature continues to rise, leading to the thermal decomposition (pyrolysis) of the organic matrix. This step is often referred to as devolatilization or distillation, and it is the key transition from a condensed phase to a reactive gas-phase environment. Complex organic compounds undergo endothermic bond cleavage, forming volatile intermediates such as hydrocarbons, alcohols, aldehydes, ketones, CO, CH_4, H_2, and tar-like compounds. The rate of devolatilization depends on the particle composition, size, heating rate, and ambient pressure. A charred residue (solid carbonaceous phase) remains after this step, which may undergo heterogeneous combustion later. The gas products formed during devolatilization are thermally and chemically reactive and provide the fuel for the subsequent gas-phase combustion.

(3) Mixing with Air (Oxidizer Diffusion)
The volatile gases evolved from the particle must now mix with oxygen from the surrounding air to form a flammable mixture. The quality and rate of mixing are determined by the turbulence level, dust cloud dispersion, and local stoichiometry. If the mixing is adequate and the local fuel–air ratio falls within the flammability limits, ignition is possible. In dust explosions, this process occurs rapidly and locally, near the ignition source, and is highly transient. Stoichiometric or near-stoichiometric pockets ignite first, followed by propagation into leaner or richer regions. The degree of premixing between volatiles and air strongly affects flame structure and combustion efficiency.

(4) Homogeneous Combustion in the Gas Phase
Once an ignitable mixture is formed, and ignition occurs, the volatilized gases combust in the gas phase through a series of chain reactions involving

radicals (H•, OH•, O•). The combustion mechanism is like that of gas-phase hydrocarbon flames, governed by thermal feedback loops, radical propagation, and chain branching reactions. This step is exothermic, releasing large amounts of energy in a very short time, generating a pressure wave and a flame front that propagates through the dust cloud. The flame temperature typically ranges between 1800 and 2400 K, depending on fuel composition and local equivalence ratio.
If oxygen is not sufficient or mixing is incomplete, soot formation, partial oxidation products, or post-flame oxidation may occur. In some cases, residual char undergoes heterogeneous combustion (surface reaction with O_2), contributing to flame propagation in later stages.

In the homogeneous combustion case, the identification of the phenomenon that controls the overall combustion rate can be done using some dimensionless numbers defined as the ratio of the characteristic times of two potentially controlling phenomena [1]. Such an identification is not straightforward, but it requires three levels of comparison to be done as the controlling step can be determined by looking at different characteristic times: both internal (t_i) and external heat transfer (t_e), pyrolysis reaction rate (t_r), evaporation (t_v), and homogeneous combustion (t_c).

The characteristic time of the internal heat transfer by thermal conduction inside a single dust particle (t_i) can be estimated as reported by Di Blasi [12]:

$$t_i = \frac{c_p \rho d^2}{\lambda} \tag{3.2}$$

where ρ and c_p are, respectively, the density and the specific heat of the solid particle and d is a characteristic length (which is, for spherical particles, the average diameter).

On the contrary, the characteristic time of the external heat transfer (t_e) can assume different expressions depending on the dominating phenomenon under consideration.

$$\text{Radiation: } t_{e,rad} = \frac{\Delta T \rho c_p d}{Q_r} \tag{3.3}$$

$$\text{Convective heating: } t_{e,conv} = \frac{\rho c_p d}{h_c} \tag{3.4}$$

$$\text{Hot-plate heating: } t_{wall} = \frac{\rho c_p d}{h_{wall}} \tag{3.5}$$

where ΔT is the temperature difference between gases and particles, Q_r is the intensity of the radiative heat flux, h_c is external heat transfer coefficient, and h_{wall} is the global heat transfer coefficient.

The typical characteristic time of the devolatilization reaction (t_r) can be defined as:

$$t_r = \frac{1}{k_p} \tag{3.6}$$

where k_p is the rate constant of a pseudo first-order pyrolysis chemical reaction; while, considering either water vaporization or low-temperature vaporizing particles the characteristic time of the evaporation can be expressed as:

$$t_v = \frac{d}{k_c} \tag{3.7}$$

where k_c is the global mass transfer coefficient.

The mechanism controlling the heat transfer process can be identified from the value of the Biot number, *Bi*, which is defined as the ratio between the characteristic times of the internal heat conduction characteristic time (t_i) and the external heat transfer characteristic time (t_e):

$$Bi = \frac{t_i}{t_e} \tag{3.8}$$

Two limiting conditions can be observed, depending on the value assumed by the Biot number:

- $Bi \ll 1$: the internal heat transfer is much faster than the external one; therefore, the heating process is controlled by the external heat transfer rate.
- $Bi \gg 1$: the internal heat transfer is much slower than the external one; thus, internal heat transfer is the controlling heating mechanism.

This classification can be considered as the "first level" of the controlling step determination for the dust combustion process. Successively, the controlling heat transfer characteristic time should be compared with the one associated with either the pyrolysis or the evaporation phenomena. Particularly, for $Bi \ll 1$, the external heat transfer (which is the slowest heat transfer process) should be compared with the pyrolysis/evaporation through the use of the Damköhler number (*Da*), which is the ratio between the characteristic times of the external heat transfer (t_e) and the pyrolysis/evaporation (t_r):

$$Da = \frac{t_e}{t_r} \tag{3.9}$$

For $Bi \gg 1$, the internal heat transfer (which is the slowest heat transfer process) should be compared with the pyrolysis/evaporation by means of the thermal Thiele number (*Th*), which is the ratio between the characteristic time of the internal heat transfer (t_i) and the pyrolysis/evaporation (t_r):

$$Th = \frac{t_i}{t_r} \tag{3.10}$$

Following such an approach, it is possible to identify the devolatilization regime under which the dust combustion phenomenon is running out. Particularly, four different devolatilization regimes can be identified within this "second level" of the controlling step determination:

- Regime I (Bi<<1 and Da>>1): the devolatilization process occurs under the external heat transfer control.
- Regime II (Bi<<1 and Da<<1) and Regime III (Bi>>1 and Th<<1): the devolatilization process occurs under the control of the pyrolysis/evaporation.
- Regime IV (Bi>>1 and Th>>1): the devolatilization process occurs under the control of internal heat transfer.

Once the devolatilization regime is identified, the step controlling the overall dust explosion phenomenon ("third level") can be determined by comparing the characteristic time of the devolatilization controlling step with the one related to the gas phase combustion (t_c) by means of the following dimensionless number [1]:

$$Pc = \frac{t_r}{t_c} = \frac{S_L}{k_p \delta} \tag{3.11}$$

where δ is the flame thickness, and S_L is the laminar burning velocity. Considering "Regime I," the external heat transfer controls the devolatilization process; therefore, to identify whether the external heat transfer or the gas phase chemical reaction controls the explosion phenomenon, it is necessary to compare two relevant characteristic times; t_e and t_c, as follows:

$$Da \cdot Pc = \frac{t_e}{t_r} \cdot \frac{t_r}{t_c} = \frac{t_e}{t_c} \tag{3.12}$$

Whenever the product $Da \cdot Pc$>>1 the external heat transfer controls, while when $Da \cdot Pc$<<1 the gas phase reaction rate controls the explosion phenomenon.

Analogously, considering "Regime IV," the internal heat transfer controls the devolatilization process and the comparison with the gas phase combustion process can be done by the means of the product:

$$Th \cdot Pc = \frac{t_i}{t_r} \cdot \frac{t_r}{t_c} = \frac{t_i}{t_c} \tag{3.13}$$

When $Th \cdot Pc$>>1 the internal heat transfer controls, while when $Th \cdot Pc$<<1 the gas phase reaction rate controls the explosion phenomenon.

In both Regimes II and III the pyrolysis/evaporation controls the devolatilization process. Therefore, Pc number directly compares the two relevant characteristic times:

$$Pc = \frac{t_r}{t_c} \tag{3.14}$$

When $Pc>>1$, the devolatilization process controls, while when $Pc<<1$, the gas-phase reaction controls.

In all cases, when the gas-phase reaction controls, the particle dimension is below a critical diameter, and the particle heating and devolatilization becomes instantaneous.

As said before, the homogeneous mechanism is also relevant for particles able to gasify totally ahead of the flame, such as sulfur particles [13] and can play a role also in conjunction with heterogeneous combustion when the boiling point of the particles is not so high [14]. In this case, the rate of the volatilization process is related to the mass transfer from the liquid surface of the particle to the gas phase.

3.2.2 Heterogeneous Mechanism

One of the most important mechanisms which characterizes the combustion of metal dusts is heterogeneous oxidation. In order to describe the main phenomena involved into this mechanism, aluminum powder combustion kinetics has been chosen as a reference. For either organic or low-temperature vaporizing particles, the main mechanism controlling the violence of a combustible dust explosion is devolatilization; on the contrary, for metal dusts, the oxidation mechanism is fundamental, and it depends on both the particles' temperature and the heterogeneous kinetics [15]. In particular, two different processes are in competition: the first is the oxygen diffusion through the oxidized layer of metal oxides (whose thickness is variable), and the second is, properly, the heterogeneous oxidation kinetics. It is known that an aluminum particle is composed of an aluminum core surrounded by a protective layer of alumina (Al_2O_3). For temperatures bounded by the melting temperatures of aluminum and alumina, the oxidation occurs in the condensed phase: in fact, oxygen diffuses through the alumina layer toward the molten core of aluminum. Then, the oxygen reacts with the Al^{3+} ions located at the boundary surface between the internal aluminum core and the external alumina layer, causing the growth of the alumina layer itself [16]. This mechanism can be described by referring to four different phenomena:

- External heat diffusion, with a characteristic time t_e.
- Internal heat diffusion, with a characteristic time t_i.
- Oxygen intra-porous diffusion, with a characteristic time $t_{d,O2}$.
- Superficial oxidation of the molten aluminum core, which corresponds to the heterogeneous combustion reaction with a characteristic time $t_{c,het}$.

With respect to the homogeneous mechanism, external and internal heat transfer characteristic times can be always calculated using equations from (3.8) to (3.11); the pyrolysis/evaporation process is substituted by the oxygen intra-porous diffusion; and, finally, the combustion reaction in the gas phase does not occur because there is only an heterogeneous oxidation of either a molten or a solid metallic core. Hence, in this case, it is possible to define two new different characteristic times:

$$t_{d,O2} = \frac{d^2}{\wp_m} \tag{3.15}$$

which is the characteristic time of the intra-porous diffusion, and:

$$t_{c,het} = \frac{1}{k_{comb} \cdot C_{met,ref}} \tag{3.16}$$

which is the characteristic time of the heterogeneous oxidation. $\wp_m$ is the diffusivity, k_{comb} is the rate constant of the second-order oxidation reaction between the adsorbed oxygen and the metal (such a reaction has been considered as the limiting step), and $C_{met,ref}$ is the reference concentration for the metal.

As previously shown for the homogeneous combustion, the mechanism controlling the heat transfer process can be identified from the value of the Biot number in order to determine the controlling step of the diffusion process, the heat transfer characteristic time (t_i or t_e) should be compared with that associated with the oxygen diffusion ($t_{d,O2}$). Particularly, for *Bi*>>1, the internal heat transfer (which is the slowest heat transfer process) should be compared with the oxygen diffusion process by means of the Lewis number (*Le*), which is equal to the ratio between the characteristic times of oxygen intra-porous diffusion ($t_{d,O2}$) and the internal heat transfer (t_i):

$$Le = \frac{t_{d,O2}}{t_i} \tag{3.17}$$

For *Bi*<<1, the external heat transfer (which is the slowest heat transfer process) should be compared with the oxygen diffusion using the inverse of the product between the Lewis (*Le*) and the Biot *(Bi)* numbers:

$$Le \cdot Bi = \frac{t_{d,O2}}{t_e} \tag{3.18}$$

As previously discussed, *Bi* can assume different expressions depending on the controlling heat transfer phenomena. Also in the case of heterogeneous combustion, four diffusion regimes can be identified:

- Regime I (*Bi*<<1 and *(Le Bi)*<<1): the diffusive process occurs under the control of the external heat transfer.
- Regime II (*Bi*<<1 and *(Le Bi)>>1*) and Regime III (*Bi*>>1 and *Le*>>1): the oxygen intra-porous diffusion is the controlling mechanism.

- Regime IV (*Bi*>>1 and *Le*<<1): the diffusion process occurs under the control of the internal heat transfer.

Once the diffusion regime is identified, the step controlling the overall dust explosion phenomenon can be determined by comparing the characteristic time of the diffusion controlling step with the characteristic time of the heterogeneous combustion ($t_{c,het}$) by means of the mass Thiele number:

$$Th = \frac{t_{d,O2}}{t_{c,het}} = \frac{d^2 k_{comb} C_{met,ref}}{\wp_m} \tag{3.19}$$

Considering "Regime I," the external heat transfer controls the diffusion process; therefore, in order to identify whether the external heat transfer or the chemical reaction controls the explosion phenomenon, it is necessary to compare two relevant characteristic times, t_e and $t_{c,het}$, as follows;

$$\frac{Th}{Le \cdot Bi} = \frac{t_e}{t_{d,O2}} \cdot \frac{t_{d,O2}}{t_{c,het}} = \frac{t_e}{t_{c,het}} \tag{3.20}$$

Whenever $Th/(Le\ Bi) >> 1$ the external heat transfer controls, while when $Th/(Le\ Bi) << 1$ the reaction rate controls the explosion phenomenon.

Analogously, considering "Regime IV," the internal heat transfer controls the diffusion process and the comparison with the combustion process can be done by means of the product:

$$\frac{Th}{Le} = \frac{t_i}{t_{d,O2}} \cdot \frac{t_{d,O2}}{t_{c,het}} = \frac{t_i}{t_{c,het}} \tag{3.21}$$

When $Th/Le >> 1$, the internal heat transfer controls, while when $Th/Le << 1$, the reaction rate controls the explosion phenomenon.

In both Regime II and III, the oxygen diffusion controls the diffusion process, and the comparison with the combustion process can be done by the means of the mass Thiele number:

$$Th = \frac{t_{d,O2}}{t_{c,het}} \tag{3.22}$$

When $Th >> 1$, the oxygen diffusion process controls, while when $Th << 1$, the heterogeneous reaction controls. Similarly to the homogeneous mechanism, when the heterogeneous reaction controls the particle dimension is below a critical diameter and the diffusion phenomena become instantaneous.

However, several other mechanisms that involve the alumina layer and that are not strictly connected to the heterogeneous oxidation mechanism. For instance, the alumina layer can be subjected to thermomechanical stresses due to aluminum melting, which could lead to the rupture of the oxide layer [15]; the same thing can occur when the alumina melting point is reached, because the oxide layer becomes thinner, and it breaks into multiple parts [17]. Depending on the temperature at which the rupture occurs, two main phenomena can compete:

- low temperature (below alumina melting point): there is a direct superficial contact between oxygen and aluminum [17] with an heterogeneous mechanism without involving the oxygen diffusion into the alumina layer;
- high temperature (equal or above alumina melting point): the oxide layer becomes so thinner that it broke into multiple parts [17]. In this case, the remained aluminum instantaneously vaporizes, and a gaseous homogeneous reaction takes phase [18].

3.3 EXPLOSION MECHANISMS

Dust explosions can be roughly classified based on both the environments in which they occur, as either confined or unconfined, and the sequential nature of the occurrence, as either primary or secondary. Understanding the distinction among the different aspects of a dust explosion is fundamental for carrying out either a proper hazard assessment or the design of effective protection systems.

3.3.1 Confined vs Unconfined Dust Explosions

A confined explosion occurs whenever a dust cloud ignites within an enclosed volume, such as a silo, duct, filter housing, pipeline, or processing equipment. In such scenarios, the combustion gases are restricted from expanding freely, leading to a rapid pressure buildup within the enclosure.

The rate of pressure rise (dP/dt) and the maximum explosion pressure (P_{max}) depend on the dust type, initial concentration, turbulence level, and degree of confinement. Typical pressure values for fully confined explosions of reactive dusts (e.g., cornstarch, aluminum) can exceed 8–10 bar, which may exceed the design pressure of industrial equipment. The confinement promotes stronger flame acceleration due to increased thermal feedback and pressure wave reflections, which amplify the combustion process.

Confined dust explosions are particularly dangerous due to their mechanical consequences, such as equipment rupture, structural collapse, and the triggering of secondary explosions through dust dispersion. Explosion severity in confined volumes is characterized by standardized parameters such as: P_{max} (Maximum Explosion Pressure), $(dP/dt)_{max}$ (Maximum Rate of Pressure Rise), K_{st} (Dust Deflagration Index, derived from $(dP/dt)_{max}$).

An unconfined dust explosion occurs in an open or semi-open environment, such as in warehouses, large rooms, or during transport, when fugitive dust accumulates in the air. In this case, the combustion gases can expand freely, and the flame propagates through the dust cloud with minimal pressure buildup. The resulting overpressures are typically much lower than in confined systems, often below 0.3–0.5

bar, unless special conditions are met (e.g., very large clouds or partial enclosures). However, the thermal and flame effects can still be devastating, especially if the fireball interacts with personnel or ignites secondary materials.

Flame propagation in unconfined dust clouds is generally slower and less turbulent, unless obstacles or partial confinement (e.g., machinery, walls) enhance flame acceleration. While unconfined explosions are less violent in terms of overpressure, they are often precursors to more destructive secondary explosions, especially in dusty environments. A small, unconfined ignition can loft accumulated dust from surfaces, forming a larger cloud that ignites in a confined location, a mechanism behind many catastrophic industrial accidents (e.g., Imperial Sugar Refinery, 2008 [19]).

3.3.2 Primary vs Secondary Explosions

A critical distinction in dust explosion scenarios is between primary and secondary explosions. Primary explosions represent the initial ignition event within a cloud of suspended dust. These typically occur in equipment such as mills, dryers, or mixers, and may be either confined or unconfined.

Secondary explosions are often far more destructive. They occur when the pressure wave or turbulence from the primary explosion resuspends dust deposits on floors, beams, or machinery, creating a new combustible cloud that ignites moments later, often in a larger, more confined space.

Secondary explosions account for the majority of fatalities and structural damage in historical dust explosion accidents, and they can involve multiple deflagration waves traveling through interconnected equipment or rooms, which may be triggered milliseconds to seconds after the primary event, making detection and suppression challenging.

Effective dust management (e.g., housekeeping, ventilation) is, therefore, essential to minimize fuel availability for secondary events.

3.4 MODELING DUST PARTICLES EXPLOSIONS

As reported in Sections 3.2.1 and 3.2.2, dust explosions must be differentiated based on their mechanism of combustion into homogeneous and heterogeneous. The same should be done whenever modeling of such phenomena is requested. It is worth noticing that while for homogeneous combustion different mathematical models can be traced in the scientific literature [20–22], for heterogeneous combustion, lots of concerns are still present [23]. Because of both this last reason and the focus of the present book (that is, both the pharma and food industries), only mathematical models referring to the homogeneous combustion process will be treated. Particularly, only the basic equations modeling a single dust particle devolatilization and combustion will be reported in the following sections, leaving to Chapter 4 (with its related sections) the task of describing how it is possible to complement the theoretical estimations of such equations for estimating experimentally the different flammability and explosibility parameters.

3.4.1 Hypotheses of the Basic Model

As a combustible dust cloud can be modeled as a group of solid particles (of different shapes, e.g., spherical, cylindrical, etc.) uniformly distributed into the space, a series of hypotheses can be stated before writing material and energy balances equations governing either the devolatilization or the combustion phenomena of each single dust particle.

Particularly, the following hypotheses can be assumed:

(i) solid dust particle of known geometry (e.g., spherical or cylindrical, which are the most relevant) surrounded by standard air (gas phase);
(ii) referring to the gas phase, negligible resistance to mass transfer and negligible diffusive fluxes in comparison with convective ones;
(iii) no secondary reactions of the volatile pyrolysis products except for combustion with oxygen in the surroundings (that is, air);
(iv) local thermal equilibrium between solid and volatiles;
(v) constant heat capacity of the solid phase;
(vi) air temperature in the surroundings of each single dust particle varying according to the phenomenon which is desired to model (e.g., simulating either a test for the determination of the maximum pressure in a 20 L sphere or the minimum ignition energy in a MIKE 3 – or further – apparatus, etc.);
(vii) pseudo-steady state assumption for the gas phase.
(viii) particles with constant volume, V_t (i.e., the porosity of the dust increases during the pyrolysis phenomenon, but this does not imply a variation of the "boundaries" of the particle itself).

3.4.2 Material Balance for a Single Solid Dust Particle

Considering a single solid dust particle, it is possible to describe the loss of material due to the pyrolysis/devolatilization step by writing a material balance equation that refers only to the solid fraction of the dust particle consumed during the process, $m_{S,r}$, (this means assuming the presence of a solid residue after the pyrolysis/devolatilization process):

$$\frac{\partial m_{S,r}}{\partial t} = -r_p \cdot V_t = -k \cdot \rho_{S,r}{}^n \cdot V_t \tag{3.23}$$

where: r_P is the rate of the pyrolysis reaction (defined as the product of the kinetic constant k and the reactive mass per unit volume of the particle (i.e., the density), $\rho_{S,r}$, expressed as $\rho_{S,r} = \frac{m_{S,r}}{V_t} = \frac{m_S}{V_t} = \frac{m_{S,0} \cdot \beta}{V_t}$, raised to the nth power); β is the residue of the particle at the end of pyrolysis (e.g., as registered in a thermogravimetric test carried out under nitrogen flux) [-].

By defining the term $\rho_{S,app} = \frac{m_S}{V_t} = \rho_S \cdot (1-\varepsilon)$, with ε, porosity of the powder particle, and $\rho_{S,app,0} = \frac{m_{S,0}}{V_t} = \rho_S \cdot (1-\varepsilon_0) \approx \rho_S$ ($\rho_{S,app,0}$ is the apparent density at the initial instant), the following expression can be obtained: $\rho_{S,r} = \rho_{S,app} - \rho_{S,app,0} \cdot \beta$.

Substituting these relations in the material balance equation and indicating the initial conditions, it is possible to obtain:

$$\begin{cases} \frac{\partial \rho_{S,r}}{\partial t} = -k \cdot \rho_{S,r}{}^{n} \\ I.C.: \rho_{S,r}(t=0) = \rho_{S,r,0} = \rho_{S,app,0} \cdot (1-\beta) \end{cases} \tag{3.24}$$

The equation can be made dimensionless by introducing the dimensionless density of the particle, $c_S = \rho_{S,r} / \rho_{S,r,0}$:

$$\begin{cases} \frac{\partial c_S}{\partial t} = -k \cdot \rho_{S,r,0}{}^{n-1} \cdot c_s{}^{n} \\ I.C.: c_s(t=0) = 1 \end{cases} \tag{3.25}$$

k is the pyrolysis kinetic constant, which can be expressed by the following modified Arrhenius equation with all the parameters obtainable by fitting a thermogravimetric analysis curve (nitrogen flux):

$$k = A \cdot exp\left[-\frac{E_a \cdot (1-\chi \cdot (1-c_S))}{RT} \right] \tag{3.26}$$

where: A was the pre-exponential factor, E_a was the activation energy and χ was the activation energy modifying factor. This last parameters takes into account the change of activation energy during the pyrolysis process due to the change in the solid matrix composition. Particularly, such a parameter can be either positive or negative depending on whether the devolatilization becomes easier or more complex as the temperature rise (and, consequently, the reaction path goes to completion).

3.4.3 Material Balance for Volatiles

During the pyrolysis/devolatilization step, a certain amount of either vapors or gases develops from the solid dust particle. Such gases go into the so-called "gas phase," which fills not only the external environment (that is, out of the dust particle boundary) but also the internal holes, which form during the process (also responsible for the porosity increase of the dust particle).

Defining the apparent density of the volatile compounds as $\rho_{V,app} = \frac{m_V}{V_t} = \frac{m_V}{V_V} \cdot \frac{V_V}{V_t} = \rho_V \cdot \varepsilon$, for the pseudo-stationary hypothesis, the material balance for volatiles can be expressed as:

$$\frac{\partial(v_r \cdot \rho_V \cdot \varepsilon)}{\partial r} = -\frac{2}{r} \cdot (v_r \cdot \rho_V \cdot \varepsilon) + k \cdot \rho_{S,r}{}^n \quad \text{Spherical coordinates} \tag{3.27}$$

$$\frac{\partial(v_x \cdot \rho_V \cdot \varepsilon)}{\partial x} = k \cdot \rho_{S,r}{}^n \quad \text{Cylindrical coordinates} \tag{3.28}$$

By introducing the massive volatiles flux exiting the single dust particle per unit surface as $v = \rho_V \cdot \varepsilon \cdot v_{r/x}$ and its ratio to $\rho_{S,r,0}$, $V = \dfrac{v}{\rho_{S,r,0}}$ (dimensionless), it is possible to obtain the final form of the material balance equation for volatiles, with its related boundary condition:

$$\begin{cases} \dfrac{\partial V}{\partial r} = -\dfrac{2}{r} \cdot V + k \cdot \rho_{S,r,0}{}^{n-1} \cdot c_s{}^n \\ B.C. \quad r = 0, \quad V = 0 \end{cases} \quad \text{Spherical coordinates} \tag{3.29}$$

$$\begin{cases} \dfrac{\partial V}{\partial x} = k \cdot \rho_{S,r,0}{}^{n-1} \cdot c_S{}^n \\ B.C. \quad x = 0, \quad V = 0 \end{cases} \quad \text{Cylindrical coordinates} \tag{3.30}$$

3.4.4 Energy Balance for the Single Particle

The energy balance equation for a single solid dust particle can be written as:

$$\rho_{S,eff} \cdot c_{p,S} \cdot \frac{\partial T}{\partial t} = -\nabla \times \left(h \cdot \vec{v} + \vec{q}\right) - \Delta H_{pyr} \cdot k \cdot \rho_{S,r}{}^n$$

where: $\rho_{S,eff} = \rho_S \cdot \left(1 - \bar{\varepsilon}\right)$ is the effective density of the solid and $\bar{\varepsilon} = 0.5$ (for simplicity of discussion) is the mean estimated porosity of the dust particle (of course, as said before, the dust particle porosity varies over time but such a variation can be considered not so significant for the energetic point of view).

The expression contains three different enthalpic contributions: the enthalpic contribution related to the convective flow of volatiles that diffused toward the outside of the dust particle, the conductive contribution related to the heating of the solid particle, and the enthalpic contribution related to the pyrolysis/devolatilization reaction of the solid. In particular, $h = \rho_{V,app} \cdot c_{p,V} \cdot T$ was the enthalpy of the volatiles per unit volume; $\vec{q} = -\bar{\lambda} \cdot \nabla T$ is the conductive heat flow, and ΔH_{pyr} is the enthalpy of the pyrolysis reaction (usually endothermic, but some exceptions, such as in the case of some charcoals, can be traced). Substituting these expressions into the energy balance equation, it is possible to obtain:

$$\rho_S \cdot \left(1 - \bar{\varepsilon}\right) \cdot c_{p,S} \cdot \frac{\partial T}{\partial t} = \bar{\lambda} \cdot \frac{\partial^2 T}{\partial r^2} + \frac{2}{r} \cdot \bar{\lambda} \cdot \frac{\partial T}{\partial r} - c_{p,V} \cdot \left[\frac{\partial}{\partial r}(v \cdot T) + \frac{2}{r}(v \cdot T)\right] - \Delta H_{pyr} \cdot k \cdot \rho_{S,r}{}^n \quad \text{Spherical coordinates} \tag{3.31}$$

$$\rho_S \cdot \left(1-\bar{\varepsilon}\right) \cdot c_{p,S} \cdot \frac{\partial T}{\partial t} = \bar{\lambda} \cdot \frac{\partial^2 T}{\partial r^2} + \frac{\bar{\lambda}}{r} \cdot \frac{\partial T}{\partial r} -$$

$$c_{p,V} \cdot \frac{1}{r} \cdot \left[\frac{\partial}{\partial r}\left(r \cdot v \cdot T\right)\right] - \Delta H_{pyr} \cdot k \cdot \rho_{S,r}{}^{n} \qquad \text{Cylindrical coordinates (3.32)}$$

The energy balance equation can be, therefore, rewritten in a pseudo-dimensionless form as:

$$\begin{cases} \dfrac{\partial T}{\partial t} = \dfrac{\lambda}{\rho_S \cdot c_{p,S}} \cdot \left(\dfrac{\partial^2 T}{\partial r^2} + \dfrac{2}{r} \cdot \dfrac{\partial T}{\partial r}\right) - \dfrac{\left(1-\beta\right) \cdot c_{p,V}}{\left(1-\bar{\varepsilon}\right) \cdot c_{p,S}} \cdot \left[\underset{t=0}{\dfrac{\partial V}{\partial r}} \cdot T + V \cdot \underset{T=T_0}{\dfrac{\partial T}{\partial r}} + \dfrac{2}{r}\left(V \cdot T\right)\right] \\ \quad - \dfrac{\Delta H_{pyr}\left(1-\beta\right)}{\left(1-\bar{\varepsilon}\right) \cdot c_{p,S}} \cdot k \cdot \rho_{S,r,0}{}^{n-1} c_s{}^{n} \\ r = 0 \qquad \dfrac{\lambda}{\rho_S \cdot c_{p,S}} \cdot \left.\dfrac{\partial T}{\partial r}\right|_{r=0} = 0 \\ \quad r = R \qquad \dfrac{\lambda}{\rho_S \cdot c_{p,S}} \cdot \left.\dfrac{\partial T}{\partial r}\right|_{r=R} = -\dfrac{h_c}{\rho_S \cdot \left(1-\bar{\varepsilon}\right) \cdot c_{p,S}} \cdot \\ \quad \left(\left.T\right|_{r=R} - T_{air}\right) - \dfrac{\varepsilon_{em} \cdot \sigma}{\rho_S \cdot \left(1-\bar{\varepsilon}\right) \cdot c_{p,S}} \cdot \left(\left.T\right|_{r=R}{}^{4} - T_{air}{}^{4}\right) \end{cases}$$

Spherical coordinates (3.33)

$$\begin{cases} \dfrac{\partial T}{\partial t} = \dfrac{\lambda}{\rho_S \cdot c_{p,S}} \cdot \left(\dfrac{\partial^2 T}{\partial r^2} + \dfrac{1}{r} \cdot \dfrac{\partial T}{\partial r}\right) - \dfrac{\left(1-\beta\right) \cdot c_{p,V}}{\left(1-\bar{\varepsilon}\right) \cdot c_{p,S}} \cdot \left[\dfrac{\partial V}{\partial r} \cdot T + V \cdot \dfrac{\partial T}{\partial r} + \dfrac{1}{r} \cdot \left(V \cdot T\right)\right] \\ \quad - \dfrac{\Delta H_{pyr} \cdot \left(1-\beta\right)}{\left(1-\bar{\varepsilon}\right) \cdot c_{p,S}} \cdot k \cdot \rho_{S,r,0}{}^{n-1} \cdot c_S{}^{n} \\ t = 0 \qquad T = T_0 \\ r = 0 \qquad \dfrac{\lambda}{\rho_S \cdot c_{p,S}} \cdot \left.\dfrac{\partial T}{\partial r}\right|_{r=0} = 0 \\ r = R \qquad \dfrac{\lambda}{\rho_S \cdot c_{p,S}} \cdot \left.\dfrac{\partial T}{\partial r}\right|_{r=R} = -\dfrac{h_c}{\rho_S \cdot \left(1-\bar{\varepsilon}\right) \cdot c_{p,S}} \cdot \left(\left.T\right|_{r=R} - T_{ext}\right) \\ \quad - \dfrac{\varepsilon_{em} \cdot \sigma}{\rho_S \cdot \left(1-\bar{\varepsilon}\right) \cdot c_{p,S}} \cdot \left(\left.T\right|_{r=R}{}^{4} - T_{ext}^{4}\right) \end{cases}$$

Cylindrical coordinates (3.34)

The initial condition refers to a temperature T_0 present at the initial instant of the simulation (and it is a function of the phenomenon, which is desired to model, as said before), while the boundary conditions refer to heat fluxes, which is zero in the "center" of the particle (r = 0) while, on the external boundary (r = R), it is given by the sum between the convective contribution and the radiating contribution expressed by the Stefan-Boltzmann law.

As it is possible to see from the boundary conditions, the knowledge of how the air temperature varies over time during the phenomenon under modeling is required to "close" the system of partial differential equations. Therefore, a suitable equation for T_{air} should be written but such an equation depends on the phenomenon under modeling [20–23], that is, how the energy is initially provided to the system.

REFERENCES

1. Di Benedetto A., Russo P., Amyotte P., Marchand N., 2010. Modelling the effect of particle size on dust explosions. *Chemical Engineering Science*, 65, 772–779.
2. Bind V.K., Roy S., Rajagopal C., 2011. CFD modelling of dust explosions: Rapid combustion in a 20 L apparatus. *Canadian Journal of Chemical Engineering*, 89, 663–670.
3. Kuai N., Huang W., Du B., Yuan J., Li Z., Gan Y., Tan J., 2013. Experiment-based investigations on the effect of ignition energy on dust explosion behaviors. *Journal of Loss Prevention in the Process Industries*, 26, 869–877.
4. Du B., Huang W., Kuai N., Yuan J., Liu L., Ren Y., 2014. Comparative study of explosion processes controlled by homogeneous and heterogeneous combustion mechanisms. *Journal of Loss Prevention in the Process Industries*, 30, 155–163.
5. Rockwell S.R., Rangwala A.S., 2013. Modeling of dust air flames. *Fire Safety Journal*, 59, 22–29.
6. Hertzberg M., Cashdollar K.L., Ng D.L., Conti R.S., 1982. Domains of flammability and thermal ignitability for pulverized coals and other dust: Particle size dependences and microscopic residue analyses. *Symposium (International) on Combustion*, 19, 1169–1180.
7. Hertzberg M., Zlochower I.A., 1990. Devolatilization rates and interparticle wave structures during the combustion of pulverized coals and poly-methlmethacrylate. *Symposium (International) on Combustion*, 23, 1247–1255.
8. Dreizin E.L., Berman C.H., Vicenzi E.P., 2000. Condensed-phase modifications in magnesium particle combustion in air. *Combustion and Flame*, 122, 30–42.
9. Dreizin E.L., 2003. Effect of phase changes on metal-particle combustion processes. *Combustion, Explosion and Shock Waves*, 39, 681–693.
10. Bind V.K., Roy S., Rajagopal C., 2012. A reaction engineering approach to modeling dust explosions. *Chemical Engineering Journal*, 207–208, 625–634.
11. Eckhoff R.K., 2003. *Dust Explosions in the Process Industries*. 3rd ed. Gulf Professional Publishing, Amsterdam & Boston.
12. Di Blasi C., 1999. Transition between regimes in the degradation of thermoplastic polymers. *Polymer Degradation and Stability*, 64, 359–367.
13. Proust C.H., 2006. A few fundamental aspects about ignition and flame propagation in dust clouds. *Journal of Loss Prevention in the Process Industries*, 19, 104–120.
14. Glassman I., 1996. *Combustion*. 3rd ed. Academic Press, New York.
15. Dufaud O., Traoré M., Perrin L., Chazelet S., Thomas D., 2010. Experimental investigation and modelling of aluminum dusts explosions in the 20 L sphere. *Journal of Loss Prevention in the Process Industries*, 23, 226–236.

16. Coker E.N., 2013. The oxidation of aluminum at high temperature studied by Thermogravimetric Analysis and Differential Scanning Calorimetry. Sandia Report. Sandia National Laboratories.
17. Escot Bocanegra P., Chauveau C., Gökalp I., 2007. Experimental studies on the burning of coated and uncoated micro and nano-sized aluminium particles. *Aerospace Science and Technology*, 11, 33–38.
18. Sun J., Dobashi R., Hirano T., 2006. Structure of flames propagating through aluminum particles cloud and combustion process of particles. *Journal of Loss Prevention in the Process Industries*, 19, 769–773.
19. Chemical Safety Board (CSB) report 2008-05-I-GA, September 2009. Last accessed 01/15/2026.
20. Copelli S., Barozzi M., Ballinari D., Lugaresi O., Moneta R., 2025. Application of critical energy density concept for minimum ignition energy determination. *Journal of Electrostatics*, 136, art. no. 104131. DOI: 10.1016/j.elstat.2025.104131.
21. Scotton M.S., Barozzi M., Derudi M., Rota R., Copelli S., 2020. Kinetic free mathematical model for the prediction of Kst values for organic dusts with arbitrary particle size distribution. *Journal of Loss Prevention in the Process Industries*, 67, art. no. 104218. DOI: 10.1016/j.jlp.2020.104218.
22. Fumagalli A., Derudi M., Rota R., Copelli S., 2016. Estimation of the deflagration index KSt for dust explosions: A review. *Journal of Loss Prevention in the Process Industries*, 44, 311–322.
23. Zhang J., Stein O.T., Luu T.D., Shamooni A., Xia Z., Luo Z., Ma L., Feng Y., Kronenburg A., 2022. Detailed modeling of aluminum particle combustion – From single particles to cloud combustion in Bunsen flames. *Chinese Journal of Aeronautics*, 35(5), 319–332.

4 Experimental Characterization of Dust Explosion Hazards

4.1 KEY PARAMETERS FOR DUST EXPLOSION CHARACTERIZATION

As dust explosions are complex combustion events depending on the physical and chemical properties of the dispersed powder, the dispersion conditions, and the confinement geometry, different parameters are typically used to quantify either the probability or the severity of a dust explosion. Such parameters are listed in Table 4.1 and detailed in the following sections.

4.1.1 Maximum Explosion Pressure (P_{MAX})

Maximum Explosion Pressure (P_{max}) is defined as the highest absolute pressure attained during the deflagration of a combustible dust–air cloud in a closed vessel under standardized test conditions. It represents the upper thermodynamic limit of pressure generation for a given dust–air system, and it is primarily governed by the extent of combustion, amount of gas generation, and adiabatic temperature rise. P_{max} characterizes the potential overpressure that may develop in fully confined equipment and, therefore, constitutes a fundamental parameter for assessing the severity of a dust explosion affecting structural resistance requirements of process vessels, enclosures, and containment systems exposed to dust explosion hazards. The test is typically performed in a 20-L spherical vessel or, less commonly, in a 1 m^3 vessel, under nearly adiabatic and constant-volume conditions.

4.1.2 Maximum Rate of Pressure Rise $(dP/dt)_{MAX}$

The Maximum Rate of Pressure Rise, denoted $(dP/dt)_{max}$, is defined as the maximum time derivative of pressure measured during the deflagration of a combustible dust–air cloud in a closed vessel under standardized conditions. It represents the highest speed at which pressure builds up during the explosion, and it is, therefore, a direct indicator of the violence and rapidity of combustion. Unlike the maximum explosion pressure P_{max}, which is dominated by thermodynamic limits, $(dP/dt)_{max}$ is primarily governed by combustion kinetics (in terms of flame propagation velocity), heat exchanged by both convection (with the gas where the dust particles are dispersed)

DOI: 10.1201/9781003645320-4

TABLE 4.1
Main Parameters for Dust Explosion Characterization

Explosion Index	EU/US Standards	Description (Risk Component)	Application
P_{max} (bar(g)) Maximum Explosion Pressure	EN 14034-1/ ASTM E1226	Maximum explosion pressure measured in a constant volume vessel. (Severity)	Design of containment, venting, and suppression systems.
$(dP/dt)_{max}$ (bar/s) Maximum rate of pressure rise	EN 14034-2/ ASTM E1226	Related to the velocity of the explosion propagation. (Severity)	As P_{max}.
K_{st} (bar m/s) Dust deflagration index	EN 14034-2/ ASTM E1226	Maximum rate of pressure rise multiplied by the cubic root of the vessel volume. (Severity)	As P_{max}.
MIE (mJ) Minimum Ignition Energy	EN 13821/ ASTM E2019	The lowest energy able to ignite the explosion propagation. (Likelihood)	Identify the degree of protection required to avoid ignition sources.
MEC (g/m^3) Minimum Explosive Concentration	EN 14034-3/ ASTM E1515	Measures the lowest dust concentration able to maintain a dust explosion. (Likelihood)	Control of dust accumulation.
LOC (volume%) Limiting Oxygen Concentration	EN 14034-4/ ASTM E2931	Minimum oxygen concentration for the flame propagation. (Likelihood)	Identify the degree of inerting.
MIT (°C) Minimum Ignition Temperature of a dust cloud	ISO/IEC 80079-20-2/ ASTM E1491	Lowest temperature leading to the explosion propagation. (Likelihood)	Control process temperature.
$(dP/dt)_{max}$ (bar/s) Maximum rate of pressure rise	EN 14034-2/ ASTM E1226	Related to the velocity of the explosion propagation. (Severity)	As P_{max}.

and conduction (within the particles' core), turbulence intensity, and dust dispersion quality. Consequently, $(dP/dt)_{max}$ is highly sensitive to (a) dust dispersion pressure, (b) dust particle size distribution, (c) vessel geometry and (d) ignition location. This sensitivity explains the larger experimental scatter commonly observed compared to P_{max}.

Measurements are typically performed in a 20-L spherical vessel (or a 1 m^3 vessel), using high-frequency pressure transducers with sufficient temporal resolution to accurately capture rapid pressure transients.

Because *$(dP/dt)_{max}$* scales with vessel size, it is not a vessel-independent parameter and must be normalized for comparison across test volumes. Its determination involves five main steps: (a) dispersion of the dust cloud into the test vessel using compressed air; (b) ignition with a standardized high-energy ignitor (typically 5–10 kJ); (c) continuous recording of pressure as a function of time; d) numerical differentiation of the pressure vs time curve; (e) identification of the maximum slope over all tested concentrations. The concentration yielding the highest *(dP/dt)* is often slightly richer than that producing P_{max}, reflecting the influence of flame acceleration and turbulence.

$(dP/dt)_{max}$ is a critical design parameter especially for explosion vent sizing (according to NFPA 68, EN 14491, etc.), explosion suppression system response times, and structural dynamic loading assessments. Anyway, it should be interpreted as a conservative, standardized severity metric; not a direct predictor of real scale pressure rises.

4.1.3 Deflagration Index (K_{ST})

The deflagration index, also denoted K_{st}, is a standardized measure of dust explosion severity that quantifies the maximum rate of pressure rise normalized for vessel size. It allows comparison of explosion violence across different experimental volumes, and it is, therefore, the principal parameter used to classify combustible dusts according to their explosion strength.

The deflagration index can be directly derived from the maximum rate of pressure rise through normalization with respect to the vessel volume; it is, therefore, defined as:

$$K_{st} = \left(\frac{dP}{dt}\right)_{max} \cdot V^{\frac{1}{3}} \tag{4.1}$$

where: $\left(\frac{dP}{dt}\right)_{max}$ is the maximum rate of pressure rise (bar·s^{-1}), and V is the volume of the test vessel (m^3). The reason behind a cubic-root volume scaling arises from similarity laws for turbulent flame propagation in confined geometries.

Measurements are typically performed in a 20-L spherical vessel (most common) or a 1 m^3 vessel for validation and scale comparison. All standards require: (a) high energy ignition (5–10 kJ) to ensure ignition-independent results; (b) testing across a range of dust concentrations; (c) identification of the concentration producing the highest *(dP/dt)*; and (d) calculation of K_{st} using the vessel volume.

Based on K_{st}, combustible dusts are classified into St classes, widely used in both EN and NFPA frameworks. Such classes are reported in Table 4.2, and they are very important for equipment design, explosion vent sizing, and risk communication.

Despite its widespread use, K_{st} has a number of limitations:

TABLE 4.2
St Classes for Severity Dust Explosion Classification

Class	K_{st} (bar·m·s^{-1})	Explosion Severity
St-0	0	Non-explosible
St-1	$0 < K_{st} \leq 200$	Weak explosion
St-2	$200 < K_{st} \leq 300$	Strong explosion
St-3	$K_{st} > 300$	Very strong explosion

(a) it is derived from laboratory-scale tests with idealized dispersion;
(b) it assumes geometric similarity that may not fully apply to complex industrial equipment;
(c) it does not account for flame acceleration due to obstacles or ducting.

For all these reasons, K_{st} should be regarded as a standardized severity index, not a direct predictor of real scale explosion pressure rate of increase.

4.1.4 Minimum Ignition Energy (MIE)

The Minimum Ignition Energy (MIE) is defined as the lowest amount of electrical energy capable of igniting a dispersed dust–air cloud (that is developing a combustion/oxidation phenomenon) under specified test conditions. It is often referred to as a measure of the ignition sensitivity (in terms of likelihood) of a combustible dust, and it is particularly relevant for assessing ignition hazards associated with electrostatic discharges, such as sparks, brush discharges, and propagating discharges.

Technically, the combustion/oxidation phenomenon cannot necessarily be sustained over time, but it must be detectable in a suitable experimental device (e.g., Hartmann tube or MIKE3 apparatus). It is important to underline that MIE is not an intrinsic thermodynamic property of a given substance (e.g., sugar or ibuprofen); rather, it is a system-dependent parameter reflecting the complex interaction among different variables such as: dust properties (resistivity, thermal conductivity, granulometry, and shape, etc.), dispersion conditions (e.g., in terms of degree of turbulence), ignition source characteristics (e.g., the presence of an inductance or not), and ambient environment.

The test apparatus typically consists of (a) a closed test chamber (cylindrical); (b) a dust dispersion system (usually based on compressed air injection); (c) a capacitor–resistor discharge circuit (allowing precise control of spark energy); (d) an adjustable electrode geometry (gap distance and orientation); and (e) an optical or pressure-based ignition detection. The dust is dispersed inside the test chamber, and it is ignited by a spark of variable energy.

The most important factors influencing MIE are:

(1) Particle size and distribution (smaller particles exhibit lower MIE).
Particle size is one of the most critical parameters. Decreasing particle size generally lowers MIE, due to both an increased specific surface area available for enhancing heat transfer between particles and surrounding gas, and a faster devolatilization (for organic dusts) or oxidation (for metallic powders) kinetics. Fine dust fractions (< 63 μm) are, therefore, usually tested to represent worst-case conditions.
(2) Moisture content (higher moisture imply higher MIE value).
Moisture acts as an inerting component, increasing the effective heat capacity of each dust particle and promoting evaporative cooling. As a result, MIE increases with increasing moisture content. Standards typically require testing of dried samples unless otherwise specified.
(3) Chemical composition.
Materials with high volatile content, low activation energy for oxidation or high combustion enthalpy tend to exhibit lower MIE values. Organic dusts generally have higher MIE than metal dusts, though exceptions exist.
(4) Dispersion uniformity and turbulence.
MIE is strongly affected by dust dispersion uniformity and turbulence level. In particular, an optimal dispersion (homogeneous cloud) minimizes MIE, while an excessive turbulence may increase heat losses and raise MIE.
(5) Environmental parameters.
Reduced oxygen concentration generally increases MIE, elevated initial temperatures decrease MIE and increased pressures generally lower MIE.

These dependencies explain because MIE values should always be reported together with test conditions, as required by EN and ASTM standards. It is important to note that MIE exhibits significant experimental scatter because results also depend on electrode geometry and discharge duration. Therefore, MIE is typically reported as an order-of-magnitude indicator, not an exact threshold. For this reason, standards recommend conservative interpretation, especially when MIE values are close to common electrostatic discharge energies (as those reported in Table 4.3).

If the MIE of a dust is lower than the expected discharge energy, ignition is considered credible, and mitigation measures (grounding, bonding, inerting, humidity control) are required.

4.1.5 Minimum Explosible Concentration (MEC)

The Minimum Explosible Concentration (MEC) is defined as the lowest concentration of combustible dust dispersed in air that is capable of supporting flame propagation and producing an explosion under specified test conditions. Below this concentration, the dust–air mixture is considered too lean to sustain combustion. MEC represents the lower explosibility limit of a dust cloud and is conceptually analogous to the Lower Flammability Limit (LFL) for flammable gases and vapors. However, unlike gaseous systems, dust clouds are inherently heterogeneous, making

TABLE 4.3
Typical Energies of Common Ignition Sources

Discharge Type	Typical Energy
Corona discharge	< 0.1 mJ
Brush discharge	1–10 mJ
Spark discharge (human)	10–30 mJ
Propagating brush discharge	> 100 mJ

MEC a strongly test-dependent and dispersion-dependent parameter. Considering the most common standards, MEC can be identified as the lowest dust concentration at which a measurable pressure rise, in a closed vessel under specified test conditions, is detected (for ASTM, a pressure rise exceeding 0.5 bar).

MEC testing is typically conducted in a closed vessel (often a 20-L sphere) and involves the following steps: (a) dispersion of a known dust mass into the vessel using compressed air; (b) ignition with a standardized ignition source (usually 1–10 kJ); (c) measurement of pressure–time history; (d) systematic reduction of dust concentration until no explosion is observed (confirmed by multiple repetitions).

From a combustion standpoint, MEC corresponds to the concentration at which the heat released by combustion balances (or exceeds): heat losses to surrounding gas, heat absorbed by unreacted particles, and expansion losses due to gas generation. Below MEC, the flame cannot sustain itself because: (a) particles spacing becomes too large; (b) radiative and conductive heat transfer between particles is insufficient; (c) the flame front quenches. Thus, MEC is fundamentally governed by heat balance and flame propagation criteria, not by ignition energy.

Several dust characteristics significantly affect MEC, such as: particle size and surface area (smaller particles exhibit lower MEC; standards, therefore, require testing of the worst-case fine fraction, often < 63 μm), moisture content (which acts on the effective dust heat capacity and promotes evaporative cooling; generally, increasing moisture content raises MEC, therefore, drying of samples is a standard practice unless otherwise specified), and chemical composition (dusts with high volatile content, high combustion enthalpy, and low ash content exhibit lower MEC values).

Moreover, MEC is highly sensitive to test conditions such as: dispersion quality (homogeneous clouds yield lower MEC), turbulence (moderate turbulence reduces MEC; excessive turbulence may increase heat losses), oxygen concentration (reduced O_2 increases MEC; below the LOC, explosion is impossible), initial temperature (higher temperatures slightly reduce MEC), and initial pressure (increased pressure generally lowers MEC). Because of this sensitivity, MEC values must always be reported together with test conditions.

MEC is a key parameter for dust hazard screening and classification, housekeeping and cleaning strategies, ventilation and enclosure design, and assessment of credible explosive atmospheres. In practice, dust clouds with concentrations above MEC are considered potentially explosive, even if such concentrations occur only transiently (e.g., during filling, discharge, or pneumatic transport).

Unfortunately, MEC exhibits several inherent limitations. In fact, it is not a sharp limit, but a probabilistic threshold and results often exhibit significant scatter due to dispersion variability. Industrial dust clouds are often non-uniform and transient; therefore, MEC should be interpreted as a conservative lower bound, not an exact concentration below which explosions are impossible.

4.1.6 Limiting Oxygen Concentration (LOC)

The Limiting Oxygen Concentration (LOC) is defined as the maximum oxygen concentration in a dust–air–inert gas mixture below which an explosion cannot occur, regardless of the dust concentration. At oxygen concentrations equal to or below the LOC, ignition of the dust cloud does not result in flame propagation or measurable pressure rise.

LOC represents the boundary between flammable and non-flammable atmospheres under oxygen-deficient conditions and it is, therefore, a fundamental parameter for the design and validation of inerting and explosion prevention systems.

Unlike ignition sensitivity parameters, LOC can be considered a prevention parameter: it defines conditions under which explosions are physically impossible rather than merely unlikely. Similarly to MEC, LOC can be identified by the highest oxygen concentration, in a mixture of oxygen, inert gas, and combustible dust, at which ignition does not result in flame propagation or a pressure rise exceeding a specified criterion (explosion is typically defined by a minimum pressure rise, e.g. ≥ 0.5 bar, consistent with other dust explosion test standards).

LOC is usually measured in a 20-L spherical vessel (or similar closed chamber) and involves:

1. Preparation of an oxygen–inert gas mixture at a defined O_2 concentration.
2. Dispersion of dust at or near the concentration producing maximum explosion severity.
3. High-energy ignition (to avoid ignition-limited effects).
4. Measurement of pressure vs time history.
5. Stepwise reduction of oxygen concentration until no explosion is observed (confirmed by repeated tests).

From a combustion perspective, LOC corresponds to the oxygen concentration at which the rate of heat release from dust oxidation is insufficient to compensate for heat losses to the gas and vessel walls, dilution by inert gas, or reduced flame temperature and radical concentration. In practice, below LOC, the adiabatic flame temperature falls below the minimum required to sustain chain reactions and flame propagation, causing the flame to extinguish.

The presence of an inert gas significantly affects LOC; moreover, different types of inert gases change LOC values according to the main mechanism through which the combustion is suppressed. For example, nitrogen acts primarily by oxygen dilution; otherwise, carbon dioxide provides both oxygen dilution and strong thermal/radiative heat absorption. As a result, $LOC_{CO2} < LOC_{N2}$; this means that CO_2 typically achieves inerting at 2–4 %v/v lower O_2 concentrations than nitrogen for the same dust.

LOC is affected by several factors such as: particle size (finer particles imply lower LOC), moisture content (higher moisture implies higher LOC), initial temperature (higher temperature implies lower LOC), and initial pressure (higher pressure implies lower LOC). For conservative design, standards recommend using worst-case conditions (dry, fine dust, elevated temperature).

Moreover, LOC is fundamentally different from other explosion characteristic parameters because it is independent of MIE (ignition sensitivity), P_{max} and K_{st}; in fact, once oxygen is below LOC, no explosion can occur. Therefore, LOC defines an absolute prevention threshold, not a severity measure.

It is also a key design parameter for inerting systems in silos, dryers, mills, and reactors, explosion prevention strategies, ATEX compliance for equipment operating under oxygen-deficient conditions, and continuous oxygen monitoring and control (in practice, safety margins are applied, and operating oxygen concentrations are kept well below LOC to account for measurement uncertainty, leaks, and process upsets).

Unfortunately, LOC application is subject to significant conceptual, experimental, and practical limitations. Misinterpretation or overreliance on LOC without appropriate safety margins and system-level analysis can result in underestimated explosion risk. Particularly, LOC is often mistakenly treated as a fixed material property, but it is a system-dependent threshold (that depends on dust particle size distribution, moisture content, chemical composition, type of inert gas (e.g., N_2 vs CO_2), temperature and pressure). As a consequence, LOC values determined under laboratory conditions may not be directly transferable to industrial processes operating under different conditions.

4.1.7 Minimum Ignition Temperature of a Dust Cloud (MIT)

The Minimum Ignition Temperature of a dust cloud (MIT or MIT_{cloud}) is defined as the lowest temperature of a hot surface or surrounding environment at which a dispersed dust–air cloud ignites under specified test conditions. Ignition is typically evidenced by flame propagation outside the heating zone or by visible combustion of the dust cloud. MIT characterizes the susceptibility of a dust cloud to ignition by hot surfaces, such as heated equipment walls, dryers and ovens, bearings, heaters, or process internals. Unlike the Minimum Ignition Energy (MIE), which addresses electrical or spark ignition, MIT concerns exclusively thermal ignition mechanisms.

MIT is typically measured using a vertical Godbert–Greenwald furnace whose standard procedure involves (a) setting the furnace to a predefined temperature, (b) injecting a fixed dust mass into the furnace, (c) observing whether ignition occurs (flame at furnace exit), and (d) repeating the test at progressively lower temperatures.

MIT is defined as the lowest furnace temperature at which ignition is observed, typically confirmed by repeated tests. To account for variability, the reported MIT is often rounded downward to the nearest 10 or 20°C, depending on the standard.

MIT reflects the balance between heat transfer from the hot environment to dust particles and heat losses due to convection, radiation, and gas expansion, versus the heat generation from oxidation reactions. Ignition occurs when particle heating is sufficiently rapid to initiate volatile release (for organic dusts), gas-phase oxidation and radical formation, and sustained flame propagation.

Because dust clouds are transient and dilute systems, ignition requires high temperatures and rapid heat transfer, explaining the relatively high MIT values.

MIT cannot be considered a constant for a given dust; in fact, it is strongly affected by several material parameters such as particle size (smaller particles imply lower MIT because finer particles heat up faster and oxidize more readily), moisture content (increased moisture content means a higher MIT because moisture delays heating due to evaporation and increases heat capacity), and chemical composition (a high volatile content implies lower MIT; otherwise, the presence of catalytic impurities or metals may reduce MIT).

MIT is a key parameter for setting maximum allowable surface temperatures of equipment, ATEX equipment category selection, and assessment of hot-surface ignition hazards in dryers, mills, and reactors.

Unfortunately, MIT exhibits a series of key limitations, which include: the strong dependence on furnace geometry and airflow, the visual ignition detection (which introduces subjectivity), and the eventuality that laboratory conditions may not represent industrial heating scenarios (in fact, transient industrial dust clouds may experience lower effective ignition temperatures due to turbulence or confinement). Therefore, MIT should be used as a screening and design parameter, not as an absolute ignition threshold.

4.1.8 Layer Ignition Temperature (LIT)

The Layer Ignition Temperature (LIT), also referred to as the Minimum Ignition Temperature of a dust layer (MIT_{layer}), is defined as the lowest temperature of a hot surface at which a settled layer of combustible dust ignites, smolders, or undergoes self-sustained combustion under specified test conditions.

Unlike dust cloud ignition, which involves transient and dilute systems, LIT addresses stationary dust deposits in contact with hot surfaces. These deposits represent one of the most critical ignition hazards in industrial installations because they can lead to smoldering combustion, delayed ignition, transition from smoldering to flaming combustion, and secondary dust explosions following dust resuspension.

LIT is experimentally determined considering the lowest surface temperature at which ignition, glowing combustion, or sustained smoldering of a dust layer occurs within a specified time. It is worth stating that smoldering is considered ignition, even in the absence of a visible flame.

The standard procedure for LIT determination involves:

- Applying a uniform dust layer of specified thickness onto the heated plate.
- Setting the plate to a constant temperature.
- Maintaining the temperature for a defined observation period.
- Observing the dust layer for detecting smoldering, glowing, charring, or flaming ignition.
- If ignition occurs, the temperature is reduced in subsequent tests. LIT is reported as the lowest plate temperature at which ignition is observed, typically rounded down to the nearest 10°C.

Typical ranges of LIT, considering a 5 mm layer, are reported in Table 4.4.

LIT values are generally significantly lower than dust cloud ignition temperatures (MIT), often by 100–300°C.

LIT is mainly governed by heat accumulation and slow oxidation mechanisms, rather than rapid flame propagation. The key processes involved include conductive heat transfer from the hot surface into the dust layer, thermal insulation by the dust layer leading to internal heat buildup, low-temperature oxidation reactions, exothermic decomposition, and char formation (for organic dusts). Once the rate of heat generation exceeds heat losses, a thermal runaway occurs, resulting in smoldering or ignition.

Layer thickness is one of the most critical parameters affecting LIT; in fact, increasing layer thickness lowers the effective LIT because thick layers act as better thermal insulators. Moreover, oxygen diffusion limitations due to increased thickness can promote smoldering rather than flaming. For some organic dusts, increasing layer thickness from 5 mm to 50 mm can reduce the ignition temperature by 50–150°C. This is the reason because standards require additional safety margins for unknown or uncontrolled dust accumulation.

LIT is strongly influenced by material characteristics such as:

- *Particle Size and Porosity.* Fine, porous dust layers exhibit lower LIT because they possess a higher surface area, which enhances oxidation reactions.

TABLE 4.4
Typical Ranges of LIT for Different Dusts

Dust Type	LIT Range [°C]
Pharmaceutical	200–300
Organic (food, wood, ...)	250–300
Polymer	300–400
Metallic	> 400

- *Moisture Content.* Moisture initially increases LIT (due to evaporative cooling) but, after drying, moisture can promote crust formation and internal heating so that nonlinear effects can be frequently observed.
- *Chemical Composition.* A dust having a high volatile content or containing catalytic metals (e.g., iron contamination) exhibits lower LIT; otherwise, a high ash content and the presence of inert fillers makes the LIT higher.

A critical feature of LIT is its time dependence; in fact, ignition may occur after minutes or hours, surface temperatures lower than LIT may still cause ignition given sufficient time, and smoldering can persist unnoticed and later can transition to a flaming combustion. For all these reasons, LIT must not be considered as a sharp threshold rather than a temperature–time criterion.

Anyway, LIT is considered one of the most important parameters for practical dust explosion prevention, and it is used for setting maximum allowable surface temperatures of equipment, ATEX equipment temperature classification, hot work and maintenance safety procedures, and housekeeping and dust layer thickness control. Under ATEX guidance, maximum surface temperatures are typically limited to 75–100°C below the LIT, accounting for uncertainty, layer thickness variations, and aging effects.

Key limitations of LIT use in industrial fields include (a) dependence on layer thickness and packing density; (b) sensitivity to surface material and roughness; (c) long test durations and subjective ignition detection; and (d) difficulty extrapolating laboratory results to aged or contaminated dust layers. As a result, LIT should be treated as a conservative design reference, not an absolute safe temperature.

4.2 OVERVIEW OF EXPERIMENTAL APPARATUS AND TECHNIQUES

The following sections describe in brief the different experimental apparatuses commonly used to determine the explosibility parameters reported in Section 4.1.

4.2.1 20 L Spherical Explosion Vessel (or Siwek 20-L Sphere)

The 20-liter spherical explosion vessel, commonly referred to as the Siwek 20-L Sphere, is one of the internationally standardized apparatuses for evaluating the explosion characteristics of combustible dusts. It is designed to ensure quasi-homogeneous dispersion, uniform ignition, and reproducible pressure-time histories, enabling the determination of maximum explosion pressure (P_{max}), maximum rate of pressure rise $(dP/dt)_{max}$ and deflagration index K_{st}, minimum explosible concentration (*MEC*), and limiting oxygen concentration (*LOC*). All these measurements are governed by international standards (see Chapter 5).

The apparatus (reported in Figure 4.1) consists of a stainless steel, pressure-rated spherical vessel with an internal volume of 20.0 ± 0.1 L. The spherical geometry ensures a uniform pressure propagation, the elimination of eventual preferential flame paths, and reproducible ignition and dispersion conditions. The sphere is designed to withstand peak pressures up to 20 bar(g) for safe measurement of aggressive dusts. A

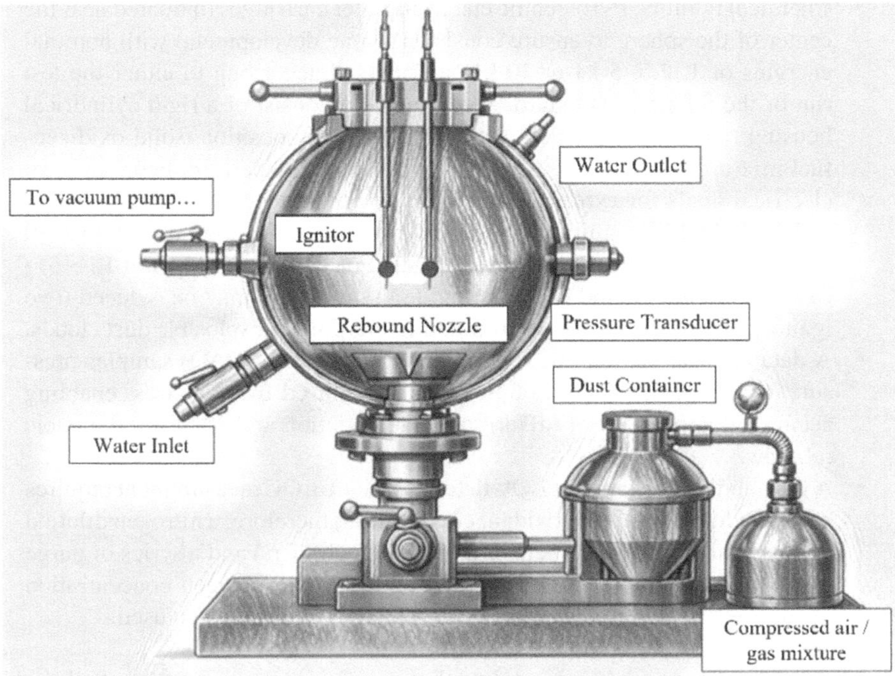

FIGURE 4.1 Sketch of the 20 L spherical explosion vessel.

hinged or flanged access port allows loading, cleaning, and installation of the dispersion cup, igniters, and instrumentation.

As one of the key features of the test method is the creation of a controlled, turbulent dust cloud, the dispersion system includes:

- A dispersion nozzle (GDI or Siwek Nozzle), mounted centrally at the bottom of the sphere. The nozzle allows high-speed injection of compressed air to disperse a metered dust charge. Typically, the dispersion pressure is of 20 bar(g) (according to ASTM E1226 and EN 14034), with an injection duration of 60 ms.
- A dust injection cup consisting of a small container, which holds the dust charge and is pneumatically emptied during the dispersion pulse. Its geometry is standardized to achieve reproducible discharge behavior.

Other elements, instrumentation, and data acquisition devices equipping the 20 L sphere are:

- A pressure transducer. Generally, fast-response piezoelectric pressure sensors (with a rise time < 1 ms) are used to record pressure–time curves, P_{max}, $(dP/dt)_{max}$. All sensors are flush mounted to avoid dust accumulation and pressure lag.

- Chemical ignitors. Pyrotechnic electric-igniter cartridges (mounted near the center of the sphere to ensure consistent flame development) with nominal energies of 1, 2.5, 5 kJ or 10 kJ can be used according to either the test run or the reference standard. They typically consist of a rigid cylindrical housing (metallic or phenolic), a pyrotechnic composition (solid oxidizer–fuel mixture), an internal electrical bridge wire or spark initiator, and two electrical leads for external triggering. In accordance with EN 14034-1/2 and ASTM E1226, ignition is provided by two chemical igniters of 5 kJ (10 kJ total), or one for more sensitive dusts. For both MEC (ASTM E1515) / LEL (EN 14034-3) and LOC testing, ignition energy may be reduced (two ignitors of 1 kJ) to avoid overstimulating marginally explosible dust clouds.
- A data acquisition system. A digital acquisition unit (DAQ) samples pressure at a frequency major of 10 kHz, as required by standards, enabling accurate computation of (dP/dt) via differentiation and K_{st} using the cubic root law.
- A gas handling system (for LOC determination). LOC measurement requires precise adjustment of the oxidant composition; therefore, a nitrogen dilution system (mass flow controllers or pre-mixed cylinders) and a series of purge cycles must be present to ensure setting the proper oxygen concentration inside the vessel. Optional oxygen/CO_2 analyzers can be also used.

The general test sequence is more or less the same for the measurement of all explosibility parameters and it can be summarized into the following steps:

- Load dust into cup.
- Seal vessel.
- Purge and set desired gas composition (LOC testing).
- Pressurize dispersion reservoir.
- At t = 0, trigger dispersion valve.
- After 60–100 ms, trigger ignition.
- Record pressure-time history until P_{max} is reached.
- Analyze $(dP/dt)_{max}$ and calculate K_{st}.
- Repeat tests ensuring the degree of reproducibility mandated by all the international standards.

4.2.2 1 m³ Spherical Explosion Vessel

The 1 m³ dust explosion vessel is the large volume standardized system for measuring dust explosion characteristics under conditions closer to industrial geometries. As the 20 L sphere, it is used to determine the following explosibility parameters: maximum explosion pressure (P_{max}), maximum rate of pressure rise $(dP/dt)_{max}$ and deflagration index K_{st}, Minimum Explosible Concentration (*MEC*), and Limiting Oxygen Concentration (*LOC*).

The 1 m³ apparatus is a spherical (or spherical–cylindrical) vessel (see Figure 4.2) with an internal free volume of 1.0 m³ (±1%), designed to withstand a

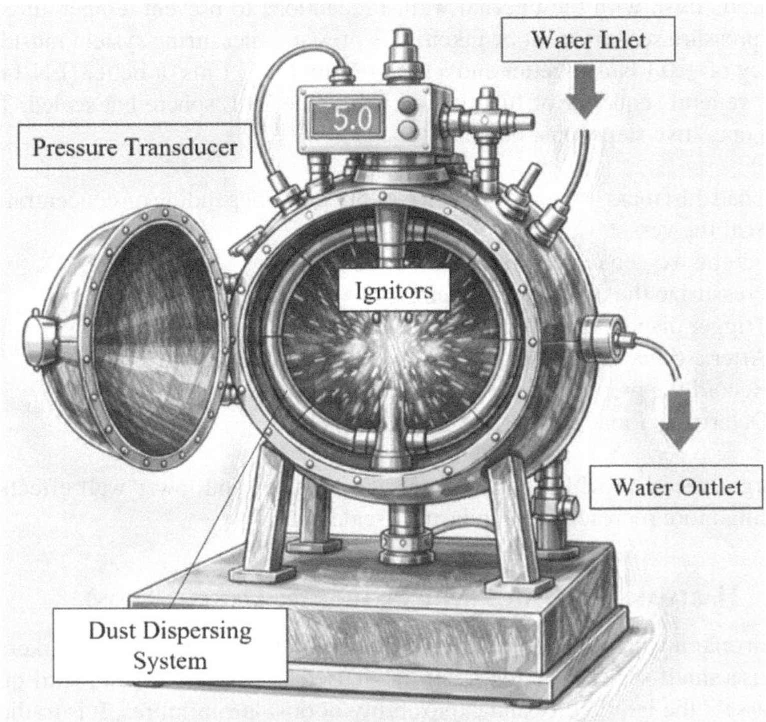

FIGURE 4.2 Sketch of a 1 m³ spherical vessel.

minimum internal pressure of 20 bar(g). The dust to be dispersed is charged into a dust container having a volume of 5.4 dm³, designed to withstand an internal overpressure of at least 20 bar(g). The dust container has an outlet at the base, through which the dust leaves the container through a fast-acting valve (designed to open in less than 10 ms) activated by a blasting cap.

For dispersing the dust, a perforated semi-circular spray pipe (dust disperser) is mounted inside the explosion vessel, concentric with its wall. For coarse, voluminous, fibrous, or poorly flowing dust samples, it may not be possible to properly discharge the dust through the dust dispersers; therefore, it may be necessary to use special dust dispersers.

The ignition source comprises two electrically activated pyrotechnical igniters, each having an ignition energy of 5 kJ. This is the nominal energy based on the mass of pyrotechnic powder in the igniters. When activated, the igniters release a dense cloud of very hot particles with little gas, then the igniters are fired by electrical fuse heads. The power supply circuit for the igniters is capable of firing the fuse heads in less than 10 ms. The two igniters must be placed at the center of the explosion vessel, firing in opposite directions (EN 14034-1).

The pressure measuring system includes at least two pressure sensors and a recording equipment. The pressure sensors must be fitted in the test vessel, with their heads flush with the internal wall. Precautions to prevent temperature effects on the pressure sensors must be taken. The pressure measuring system must have an accuracy of ± 0.1 bar or better and a time resolution of 1 ms or better (EN 14034-1).

The general sequence of functioning is like the 20 L sphere but scaled. The following operative steps must be done:

- Load dust mass (typically higher, e.g., 50–500 g depending on concentration).
- Seal the vessel.
- Set the oxygen concentration (for LOC).
- Pressurize the dispersion system.
- Trigger dispersion to form a dust cloud.
- After a timed delay (generally, 600 ms), trigger ignition.
- Record the pressure–time curve until P_{max} is reached.
- Determine Pmax, $(dP/dt)_{max}$, and calculate K_{st}.

The larger vessel provides better flame development and lower wall effects, often producing more representative industrial-scale data.

4.2.3 Hartmann Apparatus (MIE Preliminary Determination)

The Hartmann apparatus, originally developed by Hartmann and co-workers in the 1920s, is a small-scale laboratory instrument used to qualitatively and semi-quantitatively assess the ignitability and explosibility of dust–air mixtures. It is traditionally employed for either determining Minimum Ignition Energy (MIE) of dust clouds or qualitative characterization of dust explosibility (as a screening test prior to standardized measurements in a 20 L sphere). Although historically important, the Hartmann apparatus is no longer considered a fully quantitative standard for industrial safety design. Modern MIE determinations are performed according to ASTM E2019 (or EN 13821) using MIKE-3 apparatus. However, the Hartmann device remains widely described, taught, and used for preliminary hazard assessment.

A classical Hartmann apparatus (see Figure 4.3) consists of:

- A test chamber (that is, a vertical cylindrical glass or quartz tube, typically 36 cm height with a diameter of 3.6–4.0 cm, for a total volume of 400–450 mL. The transparency of the tube allows visual observation of flame propagation and dust behavior.
- A lower explosion chamber (or bomb chamber). It is a small metal chamber (usually brass or stainless steel) located at the base of the apparatus, which houses the ignition electrodes (upper part) and contains the dust dispersion cup (or powder reservoir, lower part). The lower chamber connects to the vertical tube via a flange or gasket that allows pressure release upward into the tube.
- Enclosure and Safety Shield. Because the glass tube may be ruptured during tests, the entire apparatus is usually enclosed in a safety cage or polycarbonate shield or a vented explosion-proof cabinet.

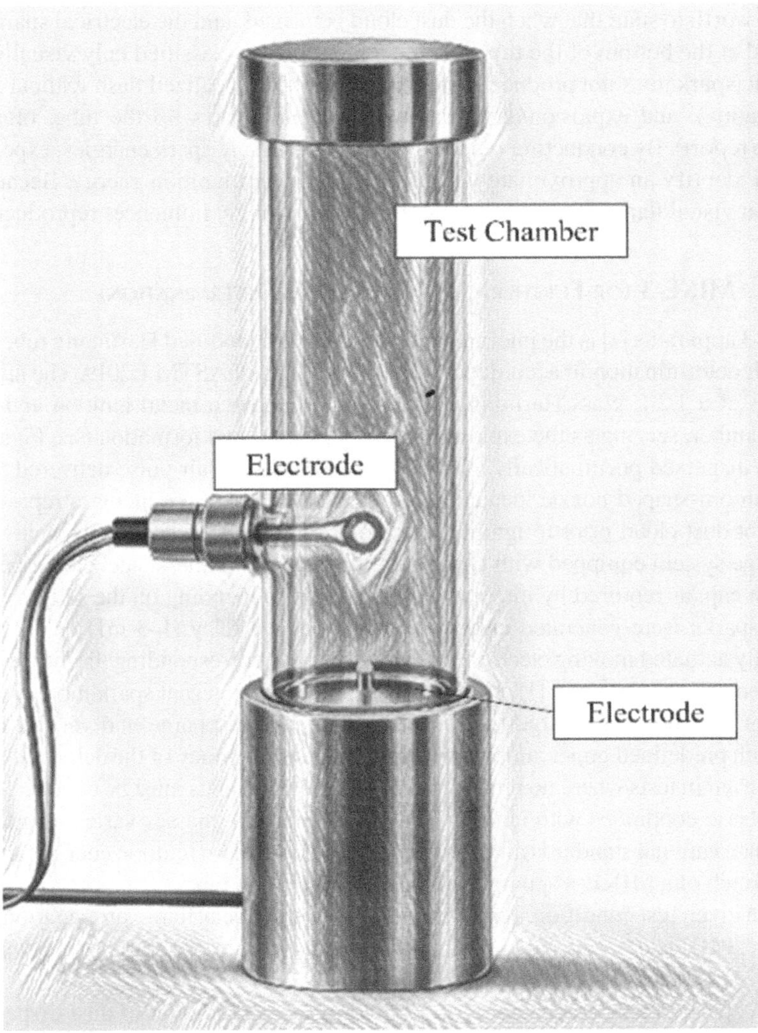

FIGURE 4.3 Sketch of a Hartmann apparatus.

Dust dispersion is achieved by a pneumatic or mechanical puff into the lower chamber: a small quantity of dust (typically 0.1–3 g) is placed in the cup; a solenoid valve injects a pulse of air (≈ 1–3 bar) into the chamber; the dust is lifted into the vertical tube, forming a dust cloud above the ignition electrodes. The dispersion is simple and not highly controlled, which limits quantitative reproducibility (because it also depends on how the powder has been loaded into the cup by the operator).

Generally, ignition is provided by two spherical electrodes spaced a fixed distance apart (typically 6–7 mm). An electrical discharge is applied to create a spark of variable energy. In classical practice, the spark energy could be varied from below 1 mJ up to 1000 mJ.

It is worth to state that when the dust cloud is formed, and the electrical spark is discharged at the bottom of the tube, the outcome of test is classified only visually as: no ignition (spark does not produce flame), weak ignition (localized flash without upward propagation), and explosion/deflagration (flame propagates up the tube, often with audible report). By conducting repeated tests at decreasing spark energies, experimenters can identify an approximate value of the minimum ignition energy. Because this relies on visual flame detection, operator training strongly influences reproducibility.

4.2.4 MIKE-3 (or Further) Apparatus (MIE Determination)

MIKE-3 apparatus [1] is the internationally recognized modified Hartmann tube system for MIE determination in accordance with EN 13821 and ASTM E2019. The apparatus consists of a 1.2-L glass Hartmann tube mounted above a metal ignition and dispersion chamber, serving as the explosion vessel for dust cloud formation (see Figure 4.4). Dust is dispersed pneumatically using a 7-bar compressed-air pulse delivered through a mushroom-shaped nozzle located at the base of the tube, producing a reproducible, turbulent dust cloud prior to ignition [1]. Ignition is achieved using a capacitive spark discharge system equipped with two 2-mm tungsten or stainless-steel electrodes set to a 6-mm gap, as required by international standards. Depending on the selected energy range, sparks were generated either via a high-voltage relay (1–3 mJ) or via a pneumatically actuated moving electrode (10 mJ–1 J), with corresponding discharge circuits documented in the manual [1]. MIKE-3 incorporates an internal spark-monitoring system that measures the charge transferred during each discharge and verifies compliance with predefined upper and lower limits, ensuring accuracy of the delivered ignition energy even in tests where no ignition occurs. All experiments must be conducted under atmospheric conditions with ignition delay times and dust masses varied as prescribed in the international standards aimed to identify the minimum ignition energy for a given dust. Sketch of a MIKE 3 apparatus (modified Hartmann tube).

For a given test condition, ignition is identified by a clear flame propagation within the tube. MIE is determined experimentally using a staircase (bracketing) method:

- A spark of a selected energy is applied to a given amount of dust dispersed in a cloud.
- If ignition occurs, the spark energy is reduced for the next trial.
- If ignition does not occur, the spark energy is increased.
- This sequence continues, also testing different amounts of dust loaded into the cup and registering if ignition occurs, until two characteristic energies are identified: E_1 (the highest energy at which no ignition occurs) and E_2 (the lowest energy at which ignition does occur).
- Finally, the statistical minimum ignition energy (E_s) is derived from E_1 and E_2 according to, for example, EN 13821.

This procedure yields the minimum spark energy capable of igniting the dust under the standardized turbulence and dispersion conditions produced by the MIKE-3 apparatus and according to a given international standard.

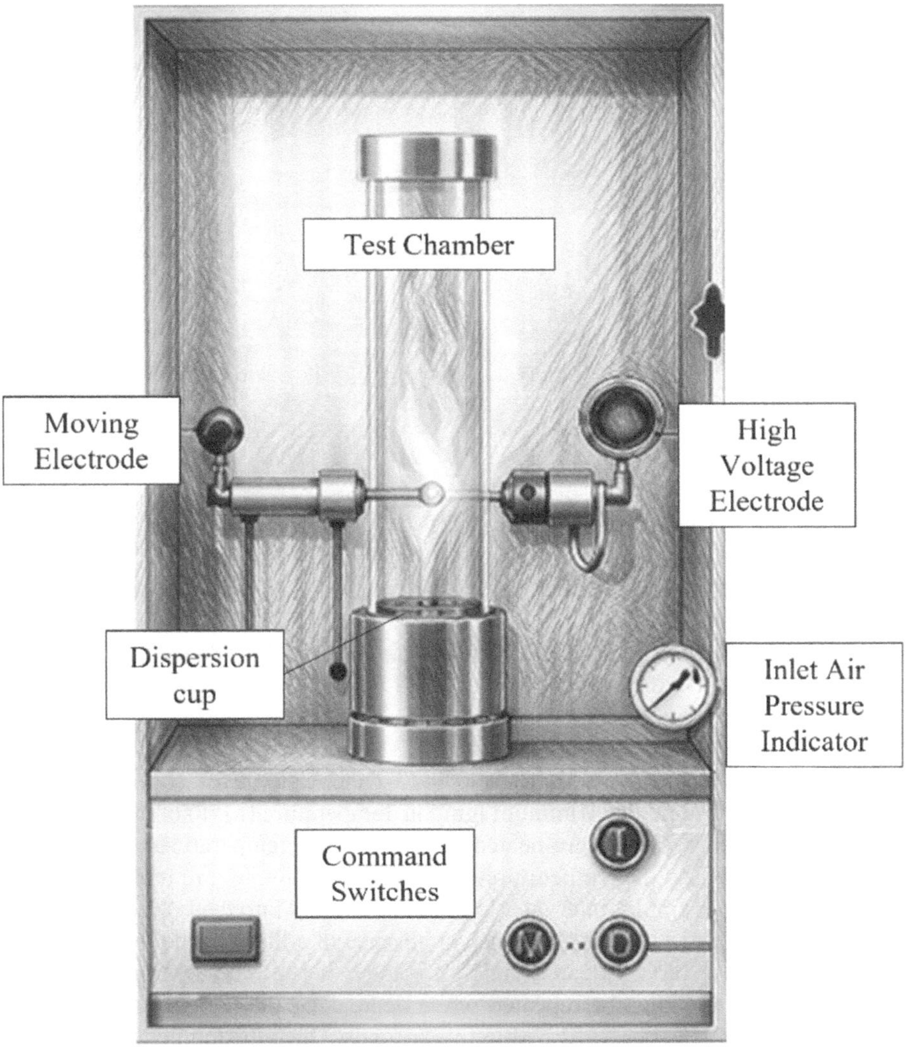

FIGURE 4.4 Sketch of a MIKE 3 apparatus (modified Hartmann tube).

4.2.5 BAM Oven (MIT Determination)

The BAM furnace, developed by the German Federal Institute for Materials Research and Testing (BAM), is a horizontally oriented tubular electric furnace made of high-temperature-resistant steel (Figure 4.5). The apparatus consists of the following main components:

- a cylindrical heated chamber (60 mm in diameter, 125 mm in length, volume = 0.35 L) closed either by a venting flap or an open outlet to the atmosphere.

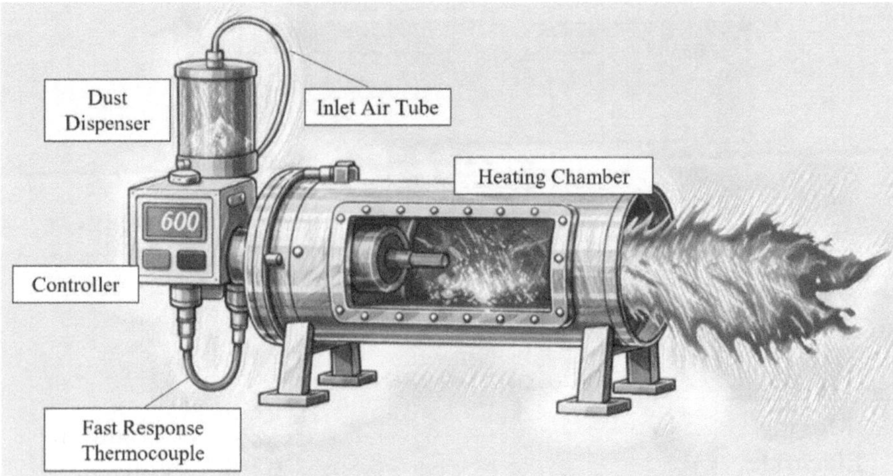

FIGURE 4.5 Sketch of a horizontal BAM furnace.

- an electric heating element and a thermocouple to measure the internal temperature.
- a disperser for generating a homogeneous dust cloud inside the chamber.
- a fast-response thermocouple connected to a recording system capable of detecting the temperature rise during ignition.

This apparatus is included in both IEC 80079-20-2 and ASTM E1491-06 as an accepted method for determining the Minimum Ignition Temperature (MIT) of dust clouds.

The furnace is typically pre-heated to a maximum temperature of 600°C and then, switched off. As the temperature decreases, dust samples are introduced using a tube at 50°C temperature intervals. The tube is inserted into the furnace, the dust is dispersed by an air jet, and the presence or absence of a flame at the rear opening is observed. The test starts with an initial dust volume of 1 mL. To determine a reliable MIT value, the test must be repeated while decreasing the temperature (10°C per step) until no flame is observed within five seconds. Once the temperature at which no ignition occurs is found, the experiment is repeated using three dust loadings (0.5, 1.0, and 2.0 mL). MIT is then reported as the lowest temperature at which flame ignition is observed for the different loadings.

Before each new test, it is essential to ensure that the chamber is completely clean, as residual burnt dust from previous runs may influence the results. Moreover, the material of construction of the chamber can have catalytic effects on ignition and must, therefore, be considered when interpreting the results.

4.2.6 Godbert–Greenwald (G–G) Furnace (MIT Determination)

The Godbert–Greenwald (G–G) furnace consists of a vertically mounted cylindrical electric furnace with an external stainless-steel casing and a hollow ceramic inner tube with an internal diameter of 365 mm (Figure 4.6).

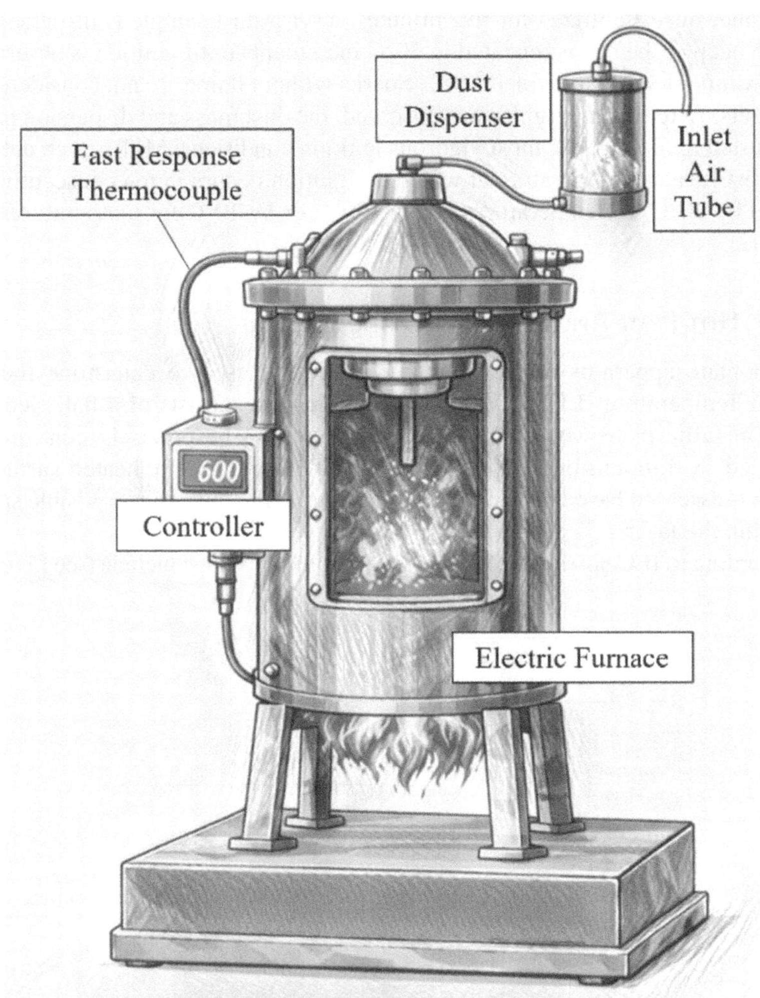

FIGURE 4.6 Sketch of a vertical Godbert–Greenwald (G–G) furnace.

The inner surface is made of refractory ceramic capable of withstanding temperatures up to 1000°C. The upper end of the furnace is connected to a glass observation elbow, which, in turn, is linked to the dust loading chamber, while the lower end remains open to allow visual confirmation of ignition using a mirror plate positioned beneath the furnace. The ceramic inner tube incorporates a spiral groove for the heating element and includes two thermocouple ports for internal temperature measurement. Dust is dispersed into the furnace by a compressed-air pulse released from a reservoir through a solenoid valve; the jet intensity is controlled by adjusting the overpressure, which is monitored using a manometer installed along the supply line.

The procedure for determining the Minimum Ignition Temperature is analogous to that employed with the BAM furnace. The furnace temperature is initially set to

500°C and, once stabilized for five minutes, a 0.1 g dust sample is dispersed. If no ignition occurs, the test is repeated in 50°C increments until ignition is observed, up to a maximum temperature of 1000°C. Sparks without flame are not considered ignition events. After the first ignition is observed, the dust mass and dispersion pressure are adjusted to identify the most vigorous ignition conditions. MIT is then defined as the lowest furnace temperature at which no ignition occurs in ten consecutive tests, reduced by 20°C for temperatures above 300°C or by 10°C for temperatures below 300°C [2].

4.2.7 Hot-Plate Apparatus (LIT Determination)

The hot-plate apparatus is the standardized device used to determine the Layer Ignition Temperature (LIT) of a combustible dust. It consists of a flat, electrically heated metallic plate whose surface temperature can be precisely controlled and monitored. A thin, uniform dust layer is placed directly on the heated surface, and ignition is assessed based on the appearance of glowing, burning, or visible smoldering within the layer.

According to IEC 80079-20-2, the hot-plate apparatus must include (see Figure 4.7):

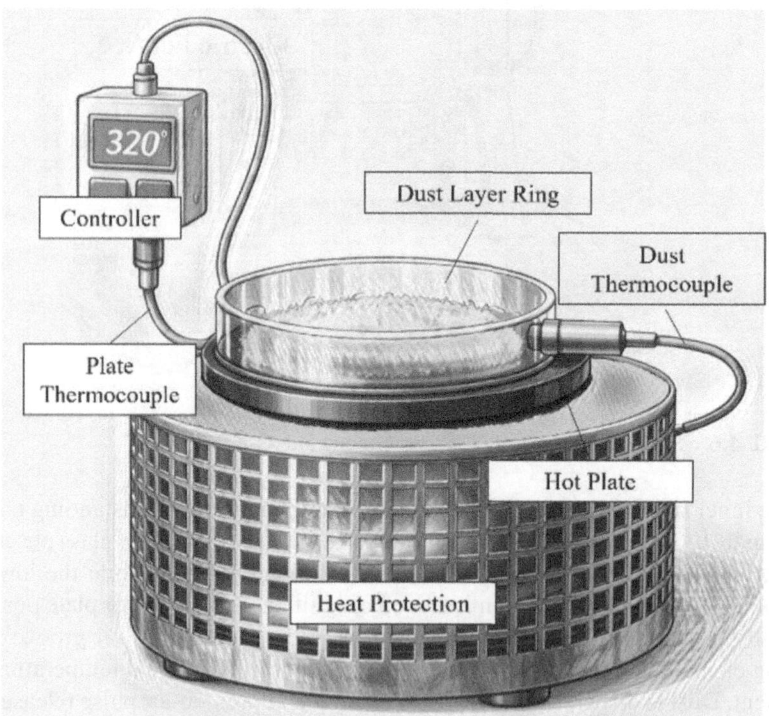

FIGURE 4.7 Hot plate apparatus for Layer Ignition Temperature determination.

- A smooth, horizontal heating surface made of metal (typically stainless steel or aluminum).
- A uniform and stable temperature field, such that the maximum temperature variation across the plate does not exceed a small tolerance specified in the standard (to ensure the entire dust layer is exposed to the intended temperature).
- A temperature-control system capable of maintaining a constant surface temperature for the duration of the test.

The plate temperature can typically be adjusted anywhere from ambient up to at least 400°C, since LIT values normally fall within this range.

IEC 80079-20-2 prescribes that a defined layer thickness (typically 5 mm, unless otherwise required) is formed using a metal ring or frame placed on the plate. The dust is then carefully leveled to achieve uniform depth and density over the heated surface; it is important to underline that the ring remains in place during heating to maintain the geometry. It should also be noticed that only dust layers of appropriate particle size and dryness, prepared according to the standard, may be tested.

During exposure, the operator monitors the dust layer for signs of ignition, defined as glowing, self-sustained smoldering, or visible flame occurring within a specified observation period (typically 30 min) after placing the dust on the hot surface (the period is defined in the standard). Other signs of ignition are also: (a) a temperature of 450°C measured in the dust layer; or (b) a temperature rise of 250 K above the temperature of the heated plate is measured in the dust layer. If none of these occur, the event counts as no ignition.

The Layer Ignition Temperature is operatively determined by:

- Setting the plate to an initial temperature chosen well below the expected LIT.
- Applying the dust layer and observing for ignition.
- Increasing the plate temperature in defined increments (standard: 10–20°C) if no ignition occurs.
- Repeating the test at each temperature until ignition is observed.

LIT is identified as the lowest plate temperature at which the dust layer ignites under the test conditions. Some laboratories perform multiple repetitions to confirm reproducibility, as recommended in the standards.

4.3 DATA ANALYSIS AND APPLICATION FOR RISK ASSESSMENT

One of the most challenging and discussed tasks of hazard identification is defining how using the values of the explosibility parameters (that is, P_{max}, K_{st}, MIE, MIT, LIT and LOC) within the analysis. The problem has already been addressed, but no explicit guidelines seem to be available in either the scientific literature or the competent authorities' platforms. This could be related to the fact that all hazard identification techniques are intrinsically "qualitative" rather than "quantitative"; even a classical Hazard and Operability Analysis (HazOp), which is the most used and reliable available technique for hazard identification, refers to whatever variable deviation without specifying any numerical boundary (e.g., High Temperature, h.T, does not require the exact computation of the temperature range; simply it is intended as a temperature higher than the normal system temperature). But, sometimes, such a "qualitative" approach could lead to an incorrect consequences identification (which subsequently affects the generation of reliable Fault Trees when the quantitative risk assessment step is carried out). In fact, given different temperature intervals, different phenomena can take place; for example, referring to a reacting system, triggering of secondary unwanted reactions, boiling or decompositions. Referring to dust hazard analysis, this concern is particularly complex to be faced because of the great variety of phenomena which can occur as a function of both dust explosibility parameters values and boundary conditions (e.g., ignition by a hot surface, ignition by an electrical spark, etc.). In the following, some guidelines on the "semiquantitative" use of the different explosibility parameters for hazard identification purposes are proposed and briefly discussed.

Considering the following explosibility parameters, that is Maximum Pressure (P_{max}), Maximum Rate of Pressure Rise ($(dP/dt)_{max}$) or Deflagration Index (K_{st}), Minimum Ignition Energy (MIE), Minimum Ignition Temperature of a dust cloud (MIT), Layer Ignition Temperature (LIT), and Limiting Oxygen Concentration (LOC), their use for Dust Hazard Analysis purposes is reported in Table 4.5.

Summarizing, in a typical dust hazard analysis, the following questions should be addressed:

- Is the dust explosible? How violent the explosion will be?
  In this case, the use of both P_{max} and K_{st} (EN 14034-1/-2, ISO/IEC 80079-20-2) to quantify explosion severity and, eventually, sizing vents, introducing suppression systems, or containment measures is required.
- How easy (or better, likely) is ignition?
  In this case, the use of MIE, MIT (cloud), and LIT is necessary to judge the dust ignition sensitivity due to the presence of sparks or hot surfaces, and to set surface-temperature limits or electrostatic controls when the basis of safety is "no effective ignition sources".
- Can the system be non-explosive by inerting?
  In this case, the use of LOC to design and control nitrogen/CO_2 inerting systems should be explored whenever an "oxidant concentration reduction" is chosen as basis of safety.

4.4 CHALLENGES IN SCALE-UP OF TESTING

4.4.1 From Laboratory Descriptors to Industrial Reality

As previously discussed, the characterization of dust explosion hazards in both pharmaceutical and food industries rely on a set of standardized laboratory-scale

TABLE 4.5
Dust Explosibility Parameters Use for Hazard Identification Purposes

Parameter	Description	General use	Dust Hazard Analysis	
			Useful checks	During HazOp
P_{max}	Maximum overpressure developed in a closed vessel for the most severe dust–air mixture under standard test conditions (ref., 20 L sphere / 1 m³ vessel; EN 14034-1, ASTM E1226, ISO/IEC 80079-20-2).	P_{max}, together with K_{st}, is a primary input for sizing explosion vents and suppression systems according to NFPA 68 and NFPA 69 (deflagration venting and prevention systems).	When the "basis of safety" is containment, the equipment is designed so its allowable pressure is greater than or equal to the expected P_{max}.	A deviation corresponding to a Very High Pressure (hh.P) should be intended as: Consequences in case of a deflagration venting failure lead to the physical explosion of the apparatus (which can be directly considered as a Top Event).
$(dP/dt)_{max}$ or K_{st}	$(dP/dt)_{max}$: maximum slope of the pressure–time curve in a closed vessel. K_{st}: volume-normalized rate of pressure rise; used to classify dusts as St-1, St-2, St-3 (EN 14034-2, ASTM E1226).	K_{st}, together with P_{max}, is a primary input for sizing explosion vents and suppression systems according to NFPA 68 and NFPA 69 (deflagration venting and prevention systems). Use K_{st} to decide whether venting alone is acceptable or active suppression/containment is needed. K_{st}-based classes (St0–St3) are widely used in hazard communication, DHA, and as input to explosion risk ranking methodologies.	Verify the correct sizing of explosion protection devices according to the reference standard. In case of incorrect sizing, before going on with, e.g., an HazOp analysis, it is necessary to substitute the deflagration venting panel.	

(Continued)

TABLE 4.5 CONTINUED
Dust Explosibility Parameters Use for Hazard Identification Purposes

			Dust Hazard Analysis	
Parameter	Description	General use	Useful checks	During HazOp
MIE	MIE is the lowest electrical spark energy (under defined conditions) that can ignite the most easily ignitable dust–air mixture (modified Hartmann tube / MIKE-3; ISO/IEC 80079-20-2, EN 13821, ASTM E2019).	IEC 60079-32-1 explicitly uses MIE as an input to decide which conductive parts must be earthed and where insulating areas must be limited. Where the basis of safety is to eliminate all credible ignition sources, a low MIE means more stringent control of electrostatics, mechanical sparks, and electrical equipment. HSG103 and EN 1127-1 treat MIE as a key indicator of ignition sensitivity for such assessments.	In DHA and electrostatic risk assessment, the energy of possible discharges (spark, brush, propagating brush) is compared with the MIE of the dust. If discharge energy ≥ MIE, those discharges must be prevented (grounding, bonding, conductive materials, humidity control). Very low-MIE powders (< 10–30 mJ) may require special handling approaches (e.g., inerting, contained transfer, limiting hybrid mixtures) instead of relying only on grounding.	Energy is not usually treated as a process variable in the strict HazOp sense, but it can and should be considered explicitly as a hazard source. This is because "energy" is not a single, directly measurable process parameter, but rather a property or potential associated with several variables. In fact, energy is implicitly covered through deviations of other variables: - Thermal energy →Temperature (More/Less temperature) - Mechanical energy → Pressure, rotating equipment speed - Electrical energy → Power supply, control systems - Chemical energy → Composition, accumulation of reactants, reaction rate - Kinetic energy → Flow velocity, momentum - Stored energy → Pressurized systems, springs, elevated masses In this sense, energy is always present in HazOp, but through its manifestations, not as a standalone variable. In some cases, especially in process safety–focused or hybrid HazOp, energy is treated as an explicit hazard parameter, for example: - High-energy reactions (runaway, decomposition) - Dust explosions (minimum ignition energy, electrostatic energy) - Electrical hazards (ESD, power failure) - Mechanical hazards (agitators, turbines, compressors) In such cases, teams may add non-standard variables like: Energy input, Ignition sources, Heat removal capacity, Static charge.

MIT	MIT (cloud) is the lowest temperature of a hot gas that will ignite a suspended dust cloud under standardized conditions (Godbert–Greenwald or BAM furnace; ISO/IEC 80079-20-2, EN ISO 80079-20-2, ASTM E1491).	MIT is used to set maximum internal temperatures of dryers, ovens, fluidized beds, and similar equipment where hot gases come into contact with dust clouds.	Industrial test labs explicitly state that process temperatures should be limited to about 2/3 of MIT (measured in degrees Centigrade) to ensure a safety margin.	A deviation corresponding to a Very High Temperature (hh.T) should be intended as: Consequences in case of protection devices failure (e.g., one shot addition of inert dusts) lead to an internal explosion, that could be referred to as a Very High Pressure (hh.P).
LIT	LIT (layer) is the lowest temperature of a hot surface that will ignite a dust layer of specified thickness (typically 5 mm) on a hot plate, under standardized conditions (hot-plate method per ISO/IEC 80079-20-2, ASTM E2021, EN 17077).	LIT is used to define allowable surface temperatures of motors, luminaires, heaters, etc., in areas where dust layers can form. Application notes and equipment data sheets explicitly use LIT to set temperature ratings for electrical equipment in dust atmospheres.	In this case, no explicit rules are stated in the standards, but typical working guidelines suggest maintaining the maximum surface temperature 75°C below the LIT [3].	A deviation corresponding to a Very High Temperature (hh.T) should be intended as: Consequences in case of protection devices failure (e.g., quench of hot surface) lead to an external explosion (which can be considered as a Top Event).
LOC	LOC is the maximum oxygen concentration at which a dust cloud cannot explode under defined conditions (EN 14034-4, ASTM E2931).	LOC is the core design parameter when the basis of safety is inerting: keeping oxygen below a value at which a dust explosion cannot occur. EN 14034-4 and ASTM E2931 define the test; NFPA 69, NFPA guidance, and DSEAR/ATEX guidance treat LOC-based inerting as a standard explosion prevention strategy.	NFPA 69 gives practical criteria on permitted oxygen concentrations relative to LOC: 1) When using continuous monitoring and control: - If LOC ≥ 5% O_2, maintain the oxygen concentration at least 2% below the lowest credible LOC. - If LOC < 5% O_2, operate at no more than 60% of LOC. 2) When oxygen is not continuously monitored or controlled with interlocks: - If LOC ≥ 7.5% O_2, maintain ≥ 4.5 vol % margin below LOC. - If LOC < 7.5% O_2, operate at no more than 40% of LOC.	A deviation corresponding to a Very High Concentration of Oxygen (hh.C_{Ox}) should be intended as: Consequences in case of protection devices failure (e.g., one shot addition of inert dusts) lead to an internal explosion, that could be referred to as a Very High Pressure (hh.P).

parameters, including maximum explosion pressure (P_{max}), maximum rate of pressure rise and derived deflagration index (K_{st}), Minimum Ignition Energy (MIE), Minimum Ignition Temperature of a dust cloud (MIT), minimum ignition temperature of a dust layer (LIT), and Limiting Oxygen Concentration (LOC). These parameters are deeply embedded in international standards, ATEX zoning practices, and engineering guidelines for explosion protection. Their widespread adoption reflects both their experimental accessibility and their apparent capacity to condense complex combustion phenomena into a manageable number of material descriptors. However, as repeatedly shown by experimental evidence at intermediate and large scales, the extrapolation of these parameters to predict either the "violence" or the probability of occurrence of industrial-scale dust explosions is far from straightforward.

The central question is not whether parameters such as K_{st}, P_{max}, or MIE are useful (of course, they clearly are), but whether they can be interpreted as intrinsic material constants capable of predicting large-scale explosion behavior without explicit consideration on dust dispersion, turbulence, and ignition dynamics. The brief discussion reported in the following sections critically examines all concerns related to the scale-up of dust explosibility parameters by explicitly linking laboratory-derived values to experimental case studies at larger scales, with particular attention to pharmaceutical and food powders.

4.4.2 K_{ST} and P_{MAX}: Necessary but Non-Sufficient Indicators of Explosion Violence

K_{st} remains the most widely used parameter for ranking dust explosion severity. Defined from the maximum rate of pressure rise measured in a standardized closed vessel (typically a 20 L sphere or 1 m³ vessel), K_{st} implicitly assumes that the cubic root scaling with volume captures the essential physics of deflagration acceleration. This assumption has proven valuable for both screening purposes and defining broad hazard classes (St-1 to St-3). However, either historical or recent experimental studies have demonstrated that K_{st} alone does not uniquely determine the violence of a dust explosion in industrial enclosures.

The large-scale silo experiments revisited by Eckhoff (2015) [4] provide a particularly striking illustration of this limitation. Wheat grain dust and soya meal dust, both tested in the 20 L sphere, showed consistently lower P_{max} and K_{st} for soya meal compared to wheat dust. Therefore, based on laboratory data only, a less violent explosion for the soya meal is expected. Yet, in 500 m³ vented silo experiments [4], the very first test with soya meal resulted in catastrophic structural failure, while the wheat dust required multiple tests and reduced vent areas to approach comparable damage levels. This discrepancy was not due to measurement error but to differences in dust cloud formation, dispersibility, and effective turbulence at ignition, none of which are captured by K_{st}.

More recent large-scale works confirm that even when K_{st} is measured accurately, it represents only a partial snapshot of the combustion process. Regis et al. [5]

showed that large-scale closed-vessel explosions with similar effective deflagration indices (K_{eff}) can exhibit markedly different pressure vs time histories, particularly during the early phase of combustion when explosion protection systems are most sensitive. Their two-parameter model demonstrated that at least two different parameters should be used to reflect the complexity of the dust explosion phenomenon (especially in its early stages): one parameter primarily reflects the turbulence intensity, while another one takes into account the intrinsic dust reactivity. K_{st} unavoidably merges these two effects into a single scalar value, masking critical differences in flame development and pressure rise dynamics.

P_{max} exhibits similar limitations. While it is often treated as an intrinsic thermodynamic property linked to dust chemistry and stoichiometry, large-scale experiments have shown that, in industrial enclosures, P_{max} is strongly influenced by heat losses, venting efficiency, flame thickness, and temporal overlap between combustion and vent flow [4]. In vented systems typical of food and pharmaceutical plants, the maximum pressure attained is often governed more by the rate of flame propagation toward vents rather than by the equilibrium pressure measured in laboratory bombs.

A key reason for the poor scalability of both K_{st} and P_{max} lies in the fundamentally different mechanisms of dust cloud generation and turbulence production at laboratory and industrial scales. Standardized tests intentionally aim to create highly dispersed, optimal dust clouds with well-defined ignition timing. In contrast, industrial dust clouds are formed by process-specific mechanisms (e.g., pneumatic conveying, mechanical agitation, gravity-driven filling, or secondary dust entrainment) that produce highly heterogeneous concentration fields and turbulence levels.

The silo experiments discussed by Eckhoff [4] highlight how dust dispersibility and agglomeration can dominate large-scale behavior. Wheat grain dust, containing fibrous material and prone to agglomeration, settled rapidly, leading to relatively homogeneous clouds of fine particles but significant mass loss prior to ignition. Soya meal, by contrast, dispersed readily and remained suspended, creating a more reactive cloud despite its lower "laboratory measured" K_{st}. Such effects cannot be reproduced in small vessels, where dispersion energy and turbulence decay times are imposed externally and largely decoupled from the material dispersibility.

In pharmaceutical contexts, this issue is even more pronounced. Many Active Pharmaceutical Ingredients (APIs) and excipients exhibit melting, softening, or decomposition near their "ignition temperatures" (it should be better to refer to an "onset temperature" of an exothermic effect detected using a combined thermogravimetric and differential scanning calorimetric approach). In laboratory tests, rapid dispersion and ignition may suppress these effects, yielding conservative K_{st} (or MIT) values. At larger scales, however, prolonged residence times, localized heating, and uneven dispersion can either inhibit or accelerate combustion, depending on process conditions. Thus, parameters measured under idealized laboratory conditions may overpredict severity in some scenarios while underpredicting it in others.

4.4.3 MIE, MIT, and LIT: Indicators of Susceptibility, Not Probability

Parameters related to ignition sensitivity, that is, MIE, MIT, and LIT, are often implicitly interpreted as measures of explosion probability. More precisely, they can be treated as "conditional thresholds under highly specific experimental configurations." MIE, for example, is measured using well-dispersed clouds and standardized spark geometries. While it provides a useful lower bound for ignition sensitivity (flammability), it does not quantify the likelihood that such an energy will be released in an industrial environment, nor does it account for the strong influence of dust concentration, turbulence, and ignition source geometry.

A pharmaceutical case study presented by Bitchikh et al. [6], regarding amoxicillin trihydrate dust, well illustrates this point. Laboratory measurements indicated a relatively high MIT (520°C) and a MIE in the range of 100–200 mJ, suggesting moderate ignition sensitivity. Nevertheless, risk assessment of the weighing room identified credible ignition scenarios arising from electrostatic discharges and hot surfaces, particularly in confined zones with dust accumulation and laminar airflow disturbances. The parameters were essential for zoning and equipment selection, but they did not, by themselves, determine the likelihood of an explosion. Instead, probability emerged from the interaction between dust release frequency, cloud persistence, ignition source presence, and housekeeping practices.

Similar considerations apply to LOC, even if such a parameter cannot be considered as a probability indicator (see considerations previously stated in Section 4.1.6). While inerting strategies in pharmaceutical reactors and food silos often rely on LOC values measured in laboratory vessels, large-scale tests have shown that stratification, air ingress, and incomplete mixing could locally exceed the LOC even when the average oxygen concentration was below the nominal limit. Thus, LOC should be viewed as a design target rather than a guaranteed safe boundary.

4.4.4 Implications for Explosion Prevention and Protection Strategies

Experimental evidence at large scale leads to a balanced interpretation of dust explosion parameters. Laboratory-scale measurements remain indispensable for hazard identification, material ranking, and preliminary design. They provide a common language for regulators, engineers, and safety practitioners, and they form the basis of standardized protection measures such as vent sizing correlations and ATEX zoning criteria.

However, the failure of K_{st}, P_{max}, and related parameters to uniquely predict large-scale explosion behavior underscores the need for context-specific analysis. Modern approaches increasingly integrate laboratory data with both intermediate and large-scale experiments, empirical correlations that explicitly include turbulence and dispersion effects, and computational fluid dynamics (CFD) models capable of simulating dust cloud formation and flame propagation. As suggested by both historical silo studies and contemporary modeling work, tailored design (rather than universal parameter-based scaling) appears to be the most robust path forward.

In the food industry, where large enclosures and highly dispersible organic dusts are common, this implies careful consideration of process-induced turbulence and secondary explosion scenarios. In the pharmaceutical industry, where powders may be more (or less) energetic but surely more sensitive to ignition and confinement, emphasis should be placed on controlling dust release, accumulation, and ignition sources, using laboratory parameters as boundary conditions rather than predictors.

In summary, dust explosion parameters measured at laboratory scale are neither meaningless nor universally predictive. They should be considered as "screening tools and design envelopes" that describe how a dust can behave under idealized conditions. The probability and violence of large-scale dust explosions emerge from the coupling of these intrinsic material tendencies with extrinsic factors such as turbulence generation, dispersion mode, confinement, and ignition location which are inherently "scale-dependent." A balanced safety philosophy, therefore, requires embedding laboratory-derived parameters within turbulence and dispersion aware frameworks, supported by large scale experiments, process knowledge, and advanced modeling. Only through such integration, the gap between laboratory characterization and industrial reality will be bridged.

REFERENCES

1. Cesana AG / REMBE Research + Technology Center, https://www.cesana-ag.ch/download/B022_075.pdf, last accessed 01/10/2026.
2. Cheremisinoff N.P., 2014. *Dust Explosion and Fire Prevention Handbook: A Guide to Good Industry Practices*. CRC Press, Taylor & Francis Group, San Diego, FL.
3. Kancar R., 2025. Understanding and preventing explosions in pharmaceutical industry. *Bartec*, November 2025. https://bartec.com/company/magazine/understanding-and-preventing-explosions-in-pharmaceutical-industry/, last accessed 01/15/2026.
4. Eckhoff R.K., 2015. Scaling of dust explosion violence from laboratory scale to full industrial scale – A challenging case history from the past. *Journal of Loss Prevention in the Process Industries*, 36, 271–280.
5. Regis C., Bauwens L., Boeck L.-R., Dorofeev S.B., 2024. Characterizing the reactivity of large-scale dust explosions with a dimensionless two-parameter model. *Proceedings of the Combustion Institute*, 40(1–4), 105280.
6. Bitchikh K., Haouche R., Bouhedda Y., Yahi Y., Smakghi N., 2025. Study of a dust explosion in a pharmaceutical production unit. *Chemical Engineering Transactions*, 117, 685–690.

5 Dust Explosion Mitigation and Risk Management

5.1 OVERVIEW OF REGULATIONS AND STANDARDS

Dust explosions represent a significant concern for process industries worldwide and each country adopts its own legislative framework to safeguard industrial and transformation activities, aiming to protect businesses, workers, and the environment. These legislative systems are shaped by national laws, decrees, and regulations that reflect each country's legal structure and its international affiliations. Therefore, while nations share a common concern regarding the potential destructiveness of dust explosions, safety regulations for process operations are uniquely structured within each jurisdiction, establishing specific requirements and obligations with the aim to mitigate, as much as possible, the occurrence of such catastrophic events. By minimizing the likelihood of dust explosions, these regulatory systems lay the groundwork for a broader range of safeguards and assurances, which can be summarized as follows:

- Protection of workers.
- Environmental protection.
- Preservation of industrial infrastructures.
- Improvement of workplaces' quality.

Failure to comply with such regulations can result also in the invalidation of insurance claims. In fact, insurance policies signed prior to an incident may not be honored if regulatory non-compliance is demonstrated in the aftermath of an explosion. The following sections will explore the main regulations and standards that influence and support the development of national laws and directives for dust explosion prevention and mitigation. The objective is to provide a general overview and concise description of the most relevant regulations, with a deliberate focus on those specifically addressing dust explosion hazards. As such, this section should not be considered a substitute for the full regulatory texts. These authoritative documents must always serve as the primary point of reference for professionals, business owners, and other stakeholders seeking reliable and comprehensive guidance on the matter.

DOI: 10.1201/9781003645320-5

5.1.1 Overview of Regulations and Standards International ISO/IEC

The International Organization for Standardization (ISO) and the International Electrotechnical Commission (IEC) are the two primary bodies responsible for defining international standards, either independently or through joint collaboration. The overarching mission of both organizations is to develop and continuously update standards that ensure the safety, quality, and interoperability of products and services on a global scale.

The IEC specifically focuses on promoting international cooperation in the standardization of electrical and electronic technologies through the publication of International Standards, Technical Specifications, Technical Reports, Publicly Available Specifications (PAS), and Guides [1]. ISO, formed from the merger of the International Federation of the National Standardizing Associations (ISA) and the United Nations Standards Coordinating Committee (UNSCC), collaborates not only with the IEC in the field of electrical and electronic standardization but also with the International Telecommunication Union (ITU) for standards development in the telecommunications sector. To date, ISO has published more than 24,000 international technical standards, complemented by Technical Specifications, Technical Reports, PAS, Guides, and International Workshop Agreements [2].

5.1.1.1 ISO/IEC 80079-20-2

The collaboration between ISO and IEC has led to the development of a range of international standards, including ISO/IEC 80079-20-2:2016, which plays a key role in the field of dust explosions. This standard consolidates and replaces previous standards IEC 61241-2-1:1994, IEC 61241-2-2:1993, and IEC 61241-2-3:1994. The purpose of ISO/IEC 80079-20-2 is to:

> describe the test methods for the identification of combustible dust and combustible dust layers in order to permit classification of areas where such materials exist for the purpose of the proper selection and installation of electrical and mechanical equipment for use in the presence of combustible dust.

The first guidance provided by the standard concerns the need to have a dust sample available, with its composition known and documented through safety data sheets. The sample must also be delivered in packaging appropriate for its intended use and labeled in accordance with relevant labeling standards. The representativeness of the sample is a fundamental requirement under the standard, as it is essential for the accurate characterization of the dust under analysis. If the sample cannot be tested in the condition in which it is received, it must be properly prepared, and such preparation must be thoroughly documented in the test report.

Before conducting the initial test, the physical structure of the dust material must be characterized using the protocol defined in the standard, with the aim of determining whether the material consists solely of dust, solely of volatiles, or a mixture of both [1]. The first test conducted to assess the material's propensity to form explosive atmospheres is the modified Hartmann tube test. This test serves as a yes/no screening procedure where the outcome of the initial test determines the subsequent steps in the characterization process, involving additional combustibility or explosibility tests. The modified Hartmann tube test represents an optional preliminary step for the user, as it is intended solely for screening purposes. Therefore, it is also acceptable to initiate the test procedure directly using the 20-liter sphere or the Godbert–Greenwald (GG) furnace, in order to characterize the material explosibility. Within the standard, both the testing equipment and the test procedures are clearly described and must be strictly followed to ensure compliance with ISO/IEC 80079-20-2 requirements for assessing the combustibility of the material. Finally, the standard provides guidance on the minimum content required in the test report, which must summarize all procedures carried out during the combustibility analysis of the material tested [1].

5.1.1.2 IEC 60079-10-2:2015

In line with ISO/IEC 80079-20-2, it is important to reference IEC 60079-10-2:2015 (Classification of areas – Explosive dust atmospheres), which addresses the classification of hazardous areas due to the presence of combustible dusts. This standard explicitly defines its relationship with ISO/IEC 80079-20-2, as well as with IEC 60079-0 (Explosive atmospheres – Part 0: Equipment – General requirements) and IEC 60079-10-1 (Explosive atmospheres – Part 10-1: Classification of areas – Explosive gas atmospheres).

The primary aim of IEC 60079-10-2 is to provide guidance for the identification and classification of work areas, where combustible dusts, fibers, or flyings may create a potential explosion hazard. It establishes the key criteria for assessing the likelihood of ignition sources in these areas. Where it is not possible to fully eliminate ignition sources, appropriate measures must be implemented to reduce the probability of forming an explosive atmosphere. If the probability cannot be reduced to a negligible level, explosion protection measures (such as isolation, suppression, or explosion venting) must be applied.

Ultimately, the goal is to assign an appropriate classification to dust hazardous areas based on the likelihood of explosive dust atmosphere formation. In addition, the standard offers recommendations for design and control strategies to prevent dust explosions, although it does not address the effects or consequences of such explosions. To support effective prevention strategies, the standard distinguishes between explosive dust clouds and accumulated dust layers, emphasizing the importance of maintaining high levels of cleanliness within facilities to minimize the presence of dust outside of process equipment.

A particularly significant point in the standard is that, while IEC 60079-10-2 does not apply to flammable gases or vapors, it may be used in conjunction with IEC 60079-10-1 for the assessment of hybrid mixture explosion risks (i.e., dust and gas/vapor combinations). Hazardous areas involving dust are classified based on the frequency and duration of explosive atmosphere formation, with the extent of the hazardous zone measured from the point of dust release. Dust layers are explicitly considered as "other sources" capable of generating explosive atmospheres. According to the likelihood of explosive dust atmosphere occurrence, the standard identifies zones as shown in Table 5.1.

According to the classification presented in Table 5.1, the following definitions apply:

- ZONE 20. This classification applies to areas where combustible dust is present continuously, for long periods, or very frequently in short durations during normal operation.
- ZONE 21. This zone refers to areas where combustible dust may occasionally be present during regular working operations, meaning that dust clouds can form under normal operating conditions.
- ZONE 22. This classification includes areas where the occurrence of explosive dust atmospheres is not likely during normal operation, and if it does occur, it will be for a short duration and with very low frequency.

Once the area classification has been completed, it becomes possible to initiate a risk assessment process aimed at identifying those zones where the use of equipment with a high Equipment Protection Level (EPL) is required, as well as those areas where it is acceptable to operate with equipment of a lower EPL. According to the definition provided in ISO/IEC 60079-0, the EPL refers to the: "level of protection assigned to equipment based on its likelihood of becoming a source of ignition, distinguishing between explosive gas atmospheres, explosive dust atmospheres, and explosive atmospheres in mines susceptible to firedamp."

5.1.1.3 IEC 60079-0:2017

Given the correlation among the various standards previously mentioned, it is also appropriate to include some notes regarding the IEC 60079-0:2017 standard (Explosive atmospheres – Part 0: Equipment – General requirements), together with its subsequent updates, including: IEC 60079-0:2017/C1:2020 IDT and IEC 60079-0:2018/A11.

The primary objective of this part of the IEC 60079 standard is to specify general requirements for the construction, testing, and marking of Ex equipment and components intended for use in areas with explosive atmospheres and standard

TABLE 5.1
Identification of Areas Based on the "Probability of Formation" of an Explosive Atmosphere

Presence of Dust	Area Classification
Continuous presence of release	20
Primary release grade	21
Secondary release degree	22

TABLE 5.2
Groups of Equipment for Use in the Presence of Explosive Atmospheres

Group	Utilization
I	Mines with risk of fireamp
II	Areas other than mines with the presence of explosive gas atmospheres
III	Areas other than mines with the presence of explosive dust atmospheres

atmospheric conditions, as defined within the standard. Equipment designed for use in explosive atmospheres is categorized into three groups, as shown in Table 5.2.

In relation to the scope of the present document, only Group III will be briefly examined, which is further subdivided based on the nature of the dust that may form the identified explosive atmosphere. Specifically, Group III is divided into: IIIA (for combustible volatile materials, as defined in IEC 60079-0), IIIB (for combustible volatile materials and non-conductive dusts, as defined in IEC 60079-0), and IIIC (for combustible volatile materials and conductive dusts, as defined in IEC 60079-0).

The devices of Group III, as indicated in the informative annex ZY of the IEC 60079-0 standard, can also be identified with an EPL rating, defined according to the IEC 60079-2 standard, as shown in Table 5.3 [3]. The EPL levels are designated as: M (for equipment intended for mines), G (for equipment intended for explosive gas atmospheres), and D (for equipment intended for explosive dust atmospheres).

In detail, regarding Group III, it is highlighted that:

- Equipment with EPL "Da" provides protection against two independent faults and can be used in Zone 20, Zone 21, and Zone 22.
- Equipment with EPL "Db" provides protection against a single fault and can be used in Zone 21 and Zone 22.

TABLE 5.3
Equipment Groups and Categories

EN IEC 60079-0		Directive 2014/34/EU		EN 60079-10-x
EPL	**Group**	**Equip Group**	**Equip Category**	**Zones**
Ma	I	I	M1	Not Applicable
Mb			M2	
Ga	II	II	1G	0
Gb			2G	1
Gc			3G	2
Da	III		1D	20
Db			2D	21
Dc			3D	22

- Equipment with EPL "Dc" provides protection for normal operation and can only be used in Zone 22.

The requirements defined by the IEC 60079-0 standard for equipment intended for use in environments with explosive atmospheres primarily focus on aspects related to the risk assessment of ignition sources generated by hot surfaces, electric arcs, electrostatic discharges, and impacts between surfaces that can produce sparks, applicable both to equipment with metallic and non-metallic enclosures [4].

5.1.1.4 ISO/IEC 80079-36

The ISO/IEC 80079-36 standard, titled "Non-electrical equipment for explosive atmospheres – Basic method and requirements," is the main reference (especially at the European level) following the withdrawal of the EN 13463 standard, for the design, construction, testing, and marking of non-electrical equipment intended for use in areas potentially affected by explosive atmospheres. Examples include pumps, gearboxes, hydraulic and pneumatic motors, and any other combination of devices used to manufacture industrial equipment.

It is important to emphasize that these are considered non-electrical equipment even though they are often fitted with an electric motor. It is necessary to define essential measures to reduce the risk of ignition intrinsic to the machine, specifically for mechanical equipment, since measures adopted for electrical equipment may not be suitable for those classified as mechanical.

Once the ignition risk has been identified, it is possible to assign the correct Equipment Protection Level to the equipment, as defined in the ISO/IEC 60079-0 standard, to prevent ignition of the explosive atmosphere. This specific section of the ISO/IEC 80079 standard instead refers only to safety aspects related to the risk of ignition intrinsic to the equipment itself and specifies design and construction requirements for equipment intended for explosive atmospheres, in compliance with all EPL for Groups I, II, and III (as better detailed in Tables 5.2 and 5.3).

The ignition risk assessment to be carried out to identify potential ignition sources during normal operation or foreseeable or rare faults of the equipment must also take into account the presence of any dust accumulations inside the equipment itself, which could ignite. The annexes of the standard provide useful and important insights regarding:

A. The methodology to confirm the EPL assigned to equipment of Group I, II, and III.
B. The procedure to implement an ignition risk analysis for assigning the EPL to the device.
C. Explanatory and practical examples of ignition risk analysis.
D. Instrumentation and test procedures to determine whether a non-conductive material can still be a potential ignition source after being charged and producing brush discharges.

E. Misuse and predictable improper use of the equipment and possible repercussions on the correct ignition risk analysis.
F. The development of possible electrostatic discharges capable of causing ignition.
G. Acceptable types of protection for non-electrical equipment.
H. The relationship between vessel volume and auto-ignition temperature for gases and liquids.
I. The relationship between EPL and ATEX zones (Table 5.3).

5.1.2 Overview of Regulations and Standards Europe

With reference to the European regulatory framework concerning dust explosion prevention, it is possible to mention the ATEX Directives and some specific EN standards. The functioning of European Directives, including the ATEX Directives, is explained in Article 288 of the Treaty on the Functioning of the European Union [5]. According to this, a Directive sets out the objectives that each Member State must achieve, while allowing individual States the freedom to determine how to achieve those objectives [6]. On the other hand, an EN standard is a technical document that can be used as a guideline to perform activities in a repeatable manner, with the interests of various stakeholders in mind, focusing on safety and quality. These standards are developed by members of the European Committee for Standardization (CEN) and the European Committee for Electrotechnical Standardization (CENELEC), with the purpose of facilitating the internal European market for goods and services and promoting its role at the global level [7].

5.1.2.1 ATEX Directive

The European Union, regarding the risk due to the presence of potentially explosive atmospheres, has adopted two health and safety directives known as ATEX 1999/92/EC (or ATEX 153) and ATEX 2014/34/EU (or ATEX 114).

- The ATEX Directive 1999/92 defines the minimum requirements for the health and safety of workers potentially exposed to the risk of explosive atmospheres.
- The ATEX Directive 2014/34 describes the responsibilities and procedures to establish conformity with European requirements for equipment and systems used in environments potentially affected by explosive atmospheres due to the presence of flammable gases or combustible dusts.

Directive 1999/92 clearly defines what can be understood as an "explosive atmosphere," which, quoting the Directive, is considered "a mixture with air under atmospheric conditions of flammable substances in the form of gas, vapor, mist, or dust in which, after ignition, combustion spreads throughout the unburned mixture." The protection of workers' health and safety starts from defining the employer's obligations, who is responsible for preventing the formation and ignition of explosive atmospheres by adopting technical and organizational measures appropriate to the type of business. In addition to prevention, the employer must also protect workers from the consequences of an explosion by implementing measures to mitigate the event.

To act accordingly, the employer must carry out a proper risk assessment of explosive atmospheres, considering the foreseeable effects. In relation to the employer's duties, it is also required that they prepare an "explosion protection document" where the identified risks, prevention and protection measures, the classification of work areas into zones with different risks of explosive atmospheres formation, and the maintenance of equipment and worker protection systems according to safety criteria are indicated. Therefore, it is necessary to demonstrate that the equipment used and the workplaces comply with current standards for working in environments at risk of explosive atmospheres formation.

Annex I of the ATEX Directive 1999/92 provides guidance on how to divide and classify areas where explosive atmospheres may form and where explosions could potentially occur. This division is based on the frequency and duration of the presence of explosive atmospheres in the environment and is completely analogous to what is indicated in Table 5.1.

Annex II provides minimum requirements to improve protection for workers potentially exposed to explosive atmospheres. These requirements include organizational measures and actual protective measures. Furthermore, it gives indications about the relationship between the different types of zones at risk of explosive atmospheres and the categories of equipment that can be used, as summarized in Table 5.4 and indicated in Directive 94/9/EC, now Directive 2014/34/EU.

Annex III, finally, shows the warning signal used to indicate the presence of areas potentially affected by explosive atmospheres, as illustrated in Figure 5.1 (Directive 1999/92/EC).

Directive 2014/34 replaced the previous Directive 94/9 due to the need for regulatory alignment with Decision 768/2008/EC and the provisions of the Treaty on the Functioning of the European Union (TFEU). Therefore, being the result of an alignment, it did not undergo significant changes compared to the previous directive, maintaining the most substantial parts unchanged. The objective of Directive 2014/34 is to ensure that products on the market, such as protective equipment and

TABLE 5.4
Relationship between ATEX Zones and Equipment Category (Directive 1999/92)

Zones for Gas, Vapor, and Fog	Zones for Dusts	Equipment Category
0	20	1
1	21	1,2
2	22	1,2,3

FIGURE 5.1 Sketch of the sign for areas potentially affected by explosive atmospheres.

control devices for environments with explosive atmospheres, offer a high level of health and safety protection while simultaneously ensuring the proper functioning of the internal market. To achieve this objective, the Directive outlines the responsibilities of the various economic operators involved in the market for protective equipment and control devices, as defined in Directive 2014/34 in Chapter I, Article 1, for use in potentially explosive atmospheres. Directive 2014/34 states that products sold in the European Union market must be marked with the CE marking, demonstrating compliance with all safety requirements established by European regulations, and it details the preliminary steps required to obtain the CE marking for each individual product.

Annex I of Directive 2014/34 contains a specific classification for equipment, such as machines, apparatus, or devices intended for energy or material handling, which, due to potential ignition sources, pose a risk of causing an explosion, as shown in Table 5.5.

Below, focusing only on aspects related to dust explosions in industrial environments, some details are provided about the equipment included in Group II, as reported in Annex II of Directive 2014/34:

TABLE 5.5
Categorization and Description of Equipment for Protection against Explosive Atmospheres (Directive 2014/34)

Apparatus Group	Use	Category	Protection Level
I	Mine	M1	Very High
		M2	High
II	ATEX	1	Very High
		2	High
		3	Normal

- Category 1. The design and construction must be such as to prevent the product from igniting an air–dust mixture even in the event of a device malfunction. The protective measures installed must also guarantee operation in the event of two independent faults, and if one protective measure fails, a second independent measure must ensure the required level of protection.
- Category 2. The design and construction must be such as to prevent the product from igniting an air–dust mixture even in the event of a foreseeable malfunction of the device.
- Category 3. The design and construction must be such that foreseeable ignition sources during normal operation do not ignite the air–dust mixture.

Annex II of Directive 2014/34 also refers to the essential requirements that must be met by equipment and protective systems intended to operate in environments with explosive atmospheres. Among these principles, there is the concept of integrated explosion safety, according to which the manufacturer must ensure that:

(1) The equipment and protective systems themselves are not a source of explosive atmospheres.
(2) Possible ignition sources of an explosive atmosphere, identified through specific analysis and properly addressed during the design phase, are avoided according to their nature, whether electrical or non-electrical.
(3) In the event of an unintended ignition of an explosive atmosphere, it is limited as much as possible within acceptable safety limits in terms of extent, duration, or pressure developed.

In general, these products must be designed and constructed, considering possible malfunctions that could compromise their functionality, to guarantee a high safety standard in hazardous situations. Fundamental to this purpose, as indicated in Directive 2014/34, Annex II, point 10.5, is the presence of correct product marking to ensure compliance with the essential safety requirements and to identify the product-specific safety characteristics. Among the most important marking references are the CE mark (see Annex II of Regulation (EC) No 765/2008) and the specific explosion protection marking, followed by other indications necessary for safe use.

Figure 5.2 shows the CE marking and the specific explosion protection marking that must be present on equipment and systems operating in explosive atmospheres, and Figure 5.3 provides an example of a complete marking.

In the design of products, to comply with the requirements of integrated explosion safety, the manufacturer is required to make an appropriate selection of the materials used, ensuring that these materials do not themselves represent an ignition source and that they maintain the functionality of the product even under hazardous conditions. Furthermore, the design and construction must be carried out in accordance with the current state of the art in explosion protection technology. Other important aspects defined by Directive 2014/34 include proper usage instructions, the sealed construction of the products, design and construction measures to prevent

FIGURE 5.2 CE marking and specific explosion protection marking.

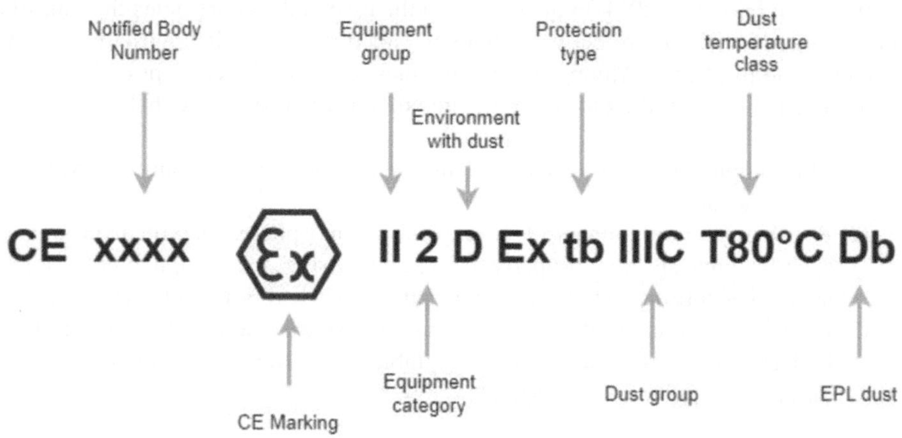

FIGURE 5.3 Example of marking for products capable of working in ATEX areas.

ignition caused by dust accumulation on the products, safe opening, protection against additional types of injury not related to explosions, overload protection for equipment, and the integration of explosion-proof protection systems. An important aspect introduced by Directive 2014/34 regarding the product conformity assessment procedure is the role of the Notified Body. This is defined as an organization recognized by the Member States of the European Union and authorized to carry out conformity assessment activities for ATEX products or systems. Product conformity may follow two paths: (a) presumption of conformity and (b) conformity assessment.

If the relevant products comply with harmonized standards, they may be presumed to meet the essential health and safety requirements set out in Directive 2014/34. In the absence of harmonized standards, the Member States determine which national standards or technical specifications must be considered by manufacturers to fulfill the essential health and safety requirements, as outlined in Annex II of Directive 2014/34. If, on the other hand, a specific product assessment must be conducted according to the provisions of Directive 2014/34, involving a conformity assessment body, or notified body, the procedure must follow the specific requirements set out in the Directive based on the relevant equipment group and category. Certifications issued by the notified body are mandatory for Category 1 and Category 2 equipment. However, manufacturers frequently issue a self-declaration of conformity for Category 3 equipment, in accordance with the provisions of Annex VIII – Internal production control. To operate as a notified body, the organization must be officially notified by the Member State in which it is based to the European Commission and to the other Member States. It is then recognized as a third-party entity authorized to assess product compliance with the provisions of Directive 2014/34. The notified body is therefore responsible for examining and testing equipment and protective systems to assess their suitability for operation in ATEX environments. It must carry out the conformity assessment in a proportionate manner, taking into account the size of the company and the technological complexity of the product, ensuring a cost-effective assessment for the manufacturer without compromising the rigor or the required level of protection necessary for product compliance (Directive 2014/34).

5.1.2.2 EN Standards

With regard to the design of protection devices against dust explosions, it is possible to refer to a series of European Standards that can be considered complementary to each other. Specifically, these are:

- EN 14491: "Dust Explosion Venting Protective System."
- EN 14797: "Explosion Venting Device."
- EN14460: "Explosion resistance equipment."
- EN 15089: "Explosion Isolation System."

EN 14491: "Dust Explosion Venting Protective System." This standard aims to provide the basic design guidelines for selecting the most appropriate explosion venting system for the protection of industrial equipment, whether isolated or interconnected, including ducts and pipelines. The venting addressed here refers to a protective device against the effects of a dust explosion occurring inside the equipment and, therefore, does not serve a prevention or extinguishing function. Its operation relies on the presence of an area of the device with lower mechanical strength than the rest of the structure, which opens and allows the release of an overpressure lower than what would be reached if the device lacked this weak point. To correctly size the venting area it is essential to have in-depth knowledge of the dust contained inside the equipment, particularly in terms of shape and particle size, as well as the conditions of the dust cloud (concentration, turbulence, and distribution) that may form and the explosiveness characteristics of the dust itself, along with some structural and geometrical parameters of the equipment. Proper sizing of the venting area is, therefore, crucial for ensuring optimal protection against dust explosions. The calculation used to determine the necessary vent area for protecting equipment from overpressures due to dust explosions must include a correction factor related to the venting efficiency factor E_f, which is specific to each piece of equipment. This factor converts the venting area A into the geometric venting area A_v according to Equation (5.1) reported below:

$$A_v = \frac{A}{E_f} \tag{5.1}$$

If the venting efficiency is maximal, i.e., equal to 1, then A_v and A are equivalent. If the efficiency is lower than 1, A_v must be greater than A to compensate for the lower efficiency and ensure adequate venting for the safe release of overpressure. An interesting consideration included in section 5.8 of the EN 14491 standard concerns the analysis of the presence of hybrid mixtures in installations, as defined in the standard itself, and how these mixtures influence the calculation of the venting area compared to the values that would apply for pure substances. Lastly, Annex D of the EN 14491 standard further explores the protection of buildings against dust explosions. The development of a dust explosion may affect not only the single piece of equipment in which it originates but also the entire building housing it. Therefore, it is important to extend the protective approach to the building's structural elements as well. The building geometry and structure, room layout, the placement of equipment within the space, and the presence of other flammable substances are among the factors that can affect the evolution of overpressure within the building. Protection measures in this context aim to limit the explosion effects to a confined portion of the total volume. For example, a primary design requirement for the building should be that even if one part is susceptible to the effects of a dust explosion, the collapse of that area should not cause a domino effect on other building components. Additionally, spaces should be designed to ensure the lowest possible risk to occupants in the event of an explosion. For this reason, it is advisable to design building sections with venting surfaces distributed evenly along the walls or roof areas. The sizing of building vents takes into account several values, including: K_{st} of the dust considered, internal surface area of the device, and maximum overpressure developed during an internal explosion with venting [9].

EN 14797: "Explosion Venting Devices." The aim of this standard is to complement the previous EN 14491 by specifying requirements for venting devices that can be placed on the market as protective systems against dust explosions for enclosures of industrial equipment. This leads to regulations concerning design, inspection, marking, documentation, and packaging. In general, the standard establishes that the venting device must not, in itself, pose an additional hazard (for instance, by projecting fragments or breaking in unintended ways when activated). This requirement must be demonstrated through explosion testing or validated modeling. Devices covered by this standard will also have specific technical characteristics that determine their performance level during operation and, consequently, their appropriate applications. These performance levels are established through dedicated laboratory testing, which includes the following:

(a) Static activation pressure. Two different testing methods can be applied, pressure test method and mechanical test method. In the pressure test method, the internal pressure is increased to a value just below the static activation pressure, followed by a linear increase over a maximum of 120 seconds. In the mechanical test method, mechanical force is gradually applied perpendicularly and statically to the device until activation occurs. The static activation pressure must be observed at both the start of the venting process and the moment in correspondence of which the venting element detaches from its support.

(b) Functionality and mechanical integrity. This parameter must be verified through explosion tests (minimum of two), performed by applying a rate of pressure rise not lower than the maximum value declared by the manufacturer for which the device was designed. Each test must be properly documented.

(c) Efficiency (when required). To determine the venting efficiency of the device, the direct comparison method may be used, which, like the other parameters, involves explosion testing. The venting efficiency of the device is calculated using Equation (5.2):

$$E_f = A_E \cdot 100 \cdot A_d^{-1} \tag{5.2}$$

where: E_f is the venting efficiency of the device; A_E is the effective venting area; A_d is the geometric area of the venting device.

In this equation, the value of A_E can be determined through direct comparison of explosion test results performed on the venting devices in question and on reference venting devices with a specific mass less than 0.5 kg·m^{-2} [10].

EN14460: "Explosion resistance equipment." This standard specifies the requirements for equipment that must be able to withstand an internal explosion without rupturing and without creating further external risks. As shown in Figure 5.4, there are two types of explosion-resistant equipment.

Explosion pressure-resistant equipment is designed to withstand explosions without undergoing permanent deformation and without creating additional hazards to

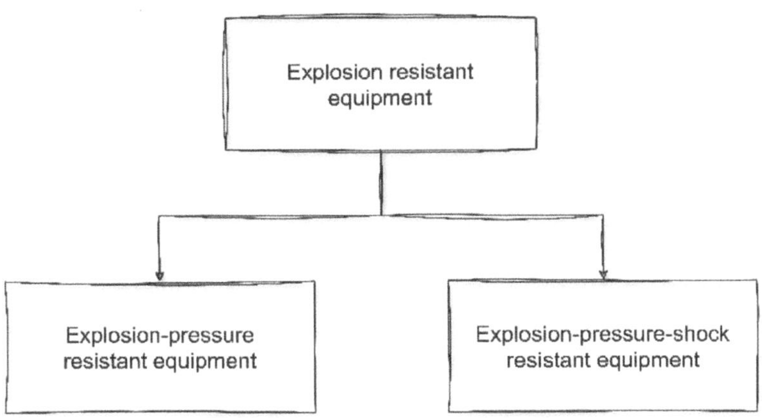

FIGURE 5.4 Explosion resistant equipment [11].

the external environment. Their design approach is similar to the guidelines set out in standards EN 13445-1 through EN 13445-6. Explosion pressure shock-resistant equipment, on the other hand, may undergo permanent deformation, provided that no hazardous conditions are created externally. This standard provides a detailed framework for the design of such equipment. Explosion pressure shock-resistant equipment must be designed by taking into account the geometry, functionality, and design pressure characteristics, with particular attention paid to the construction materials in terms of product safety. The specified design pressure for the equipment must not be lower than the maximum pressure that may occur within the equipment during an explosion, in order to ensure adequate containment. Likewise, resistance to increased temperature in the event of an explosion must also be considered. The determination of the design pressure p must take into account the initial pressure p_v only if it is lower than 90 kPa or higher than 110 kPa, and it must be calculated in accordance with Equation (5.3).

$$p = \frac{p_{\max,0} \cdot p_v}{p_0} - p_0 \tag{5.3}$$

EN 15089: "Explosion Isolation Systems." This standard provides guidelines regarding the general requirements that must be met by active and passive explosion isolation systems, such as protection valves, suppression barriers, rotary valves, and diverters. These components are specifically designed to prevent the propagation of overpressure waves or flames through pipes or ducts from the point of explosion origin to other parts of the apparatus or plant. The standard specifies that, for an explosion isolation system to function effectively, it must be capable of detecting and extinguishing or blocking the flame front, which is responsible for the propagation of the explosion. It also provides guidance on the system components that allow for detection activities and communication between the various system elements and the operator. The standard highlights the importance of conducting a thorough analysis of the conditions under which the isolation system is installed and its specific characteristics. This includes evaluating the type of connections with other devices and the nature of the process in which the system is integrated.

To experimentally test the effectiveness of an explosion isolation system, each isolation device must undergo the test procedures outlined in Table 5.6, and specific, detailed test reports must be issued, as defined by EN 15089.

The design of explosion isolation devices represents the final step in applying specific engineering methods that enable the manufacturer to determine and define a series of installation and operational characteristics of the devices themselves. These methods (tailored for passive isolation valves, active isolation valves, and suppression barriers) can be divided into two distinct categories: methods derived from experimental data and, methods based on mathematical analyses capable of simulating explosive processes [13].

TABLE 5.6
Type of Modules as a Function of Type of Isolation Device [13]

Type of Isolation Device	Module A: Explosion Resistance Test	Module B: Flame Transmission Test	Module C: Functional Test
Passive isolation valve	X	X	X
Explosion proof interlocked double valve arrengment	X	X	X
Active isolation valves	X	X	X
Extinghishing barriers	–	–	X
Rotary valves	X	X	X

5.1.3 Overview of Regulations and Standards North America: NFPA standards

The National Fire Protection Association (NFPA) is a non-profit organization based in Massachusetts (the US), whose mission is to address safety issues by prioritizing people over profit. Specifically, NFPA operates through the publication of codes and standards, conducting research, providing professional training, and promoting public education to eliminate fatalities, injuries, and material or economic losses resulting from events such as fires, electrical incidents, and related hazards. Currently, the NFPA involves more than 10,000 professional volunteers who make up over 250 technical committees tasked with developing and publishing standards in the interest of all stakeholders. Within the framework of US regulations on dust explosion prevention and control, NFPA standards serve as a fundamental support for the implementation of the International Fire Code (IFC), which, along with other regulatory codes, forms the normative basis for construction and the issuance of work permits in industrial facilities [4].

5.1.3.1 NFPA 660

Starting from December 2024, the US dust explosion regulations underwent a significant reorganization, leading to the creation of NFPA 660. This standard is to be considered as a carefully structured combination of existing regulations, providing industries with a tool that consolidates various standards related to dust explosion management, fire prevention, and best practices for protecting health, the environment, and facilities from the consequences of dust explosions [14].

One of the first and most evident innovations introduced by NFPA 660, as anticipated, is the grouping of several pre-existing NFPA standards, including:

- NFPA 61: Standard for Prevention of Fires and Dust Explosions in Agricultural and Feed Processing Facilities.
- NFPA 655: Standard for Prevention of Sulfur Fires and Explosions.

- NFPA 484: Standard for Combustible Metals.
- NFPA 652: Standard Fundamentals of Combustible Dust.
- NFPA 654: Standard for the Prevention of Fire and Dust Explosions from Manufacturing, Processing, and Handling of Combustible Particulate Solids.
- NFPA 664: Standard for the Prevention of Fires and Explosions in Wood Processing and Woodworking Facilities.

This series of standards regulated the industrial sector in terms of prevention and protection against dust explosions. The primary reference document was NFPA 652, which focused on providing general guidelines on how to recognize and identify dust explosion hazards and how to control them, while the other NFPA standards (also called "commodity-specific" or "industry-specific" standards) were more dedicated to specific industrial sectors or occupational activities. The direct relationship between these standards and the transposition into NFPA 660 is reported in Table 5.7.

The objective of the NFPA 660 standard is, therefore, to provide a definitive and clear framework containing the technical fundamentals of dust explosions and the essential information needed to manage the risk of such scenarios across various industrial sectors affected by dust explosions, including agriculture, food processing, metal and woodworking, coal processing, and chemical industries.

The main features that set the NFPA 660 standard apart from the individual standards it consolidates can be outlined as follows:

- Risk Assessment and Hazard Identification. NFPA 660 introduces comprehensive approaches for carrying out detailed dust hazard analyses, implementing explosion prevention mechanisms, assessing risks, and recognizing potential dust-related dangers, including electrical components within industrial settings.
- Preventive Measures and Controls. The standard specifies preventive strategies and engineering solutions, such as dust collection infrastructure and dust extraction equipment, aimed at controlling dust accumulation and

TABLE 5.7
Transposition of Dust Explosion NFPA Standards in NFPA 660

NFPA Standard	Transposition in NFPA 660
NFPA 652	From Cap. 1 to Cap. 10
Reserved for Future implementation	From Cap. 11 to Cap. 20
NFPA 61	Cap. 21
NFPA 484	Cap. 22
NFPA 655	Cap. 23
NFPA 664	Cap. 24
NFPA 654	Cap. 25

minimizing ignition sources, including static discharge and other potential triggers.

- Housekeeping and Maintenance Practices. Emphasizing the importance of routine cleaning and upkeep, NFPA 660 advocates for practices that reduce dust buildup, maintain dust control systems, and ensure the reliable performance of safety devices.
- Training and Education. The standard highlights the critical need for thorough training and education of personnel who handle, process, or manage combustible materials, with particular attention to recognizing and mitigating ignition sources like static electricity.
- Emergency Response and Preparedness. NFPA 660 offers detailed guidance on creating emergency response strategies and procedures designed to effectively lessen the impact of dust-related fires and explosions [14].

The NFPA 660 standard is organized into ten main chapters (Chapters 1 to 10), followed by an additional set of ten "Reserved" chapters (Chapters 11 to 20), which are reserved for future developments on emerging topics such as Fire Protection, Additive Manufacturing, and Nanomaterials. Beyond Chapter 20, which is the last reserved chapter for future integration, there are five more chapters (Chapters 21 to 25). Within the completed chapters, there may also be specific "Reserved" sections for future updates, for example, Section 5.4.2 titled "Determination of Flash Fire Potential." Moreover, NFPA 660 contains 26 annexes (Annex A to Annex Z) that offer supplementary detailed guidance on specialized subjects.

Following the initial three chapters dedicated to administrative introductions, references, and definitions, Chapter 4, "General Requirements," presents a range of elements and guidance aimed at assisting plant owners in achieving objectives related to the identification, evaluation, management, and communication of combustible dust explosion risks. These objectives are to be achieved through adequate protection of workers from the effects of dust-related explosions and fires, mitigating and preventing such incidents by ensuring proper implementation of prescriptive requirements and risk management plans in line with NFPA 660.

Chapter 5, "Hazard Identification," delves deeper into risk recognition, first emphasizing that if the explosive or combustible properties of a dust are unknown, appropriate testing must be conducted to gather necessary data to determine the potential dust explosion hazard. Additionally, specific risk assessments must be performed for self-ignition, water reactivity, and chemical reactivity. Starting from Section 5.4, the standard outlines detailed procedures for conducting combustibility and explosivity tests on dust samples and for obtaining representative samples.

Chapter 6, "Performance-Based Design Option," provides an alternative tool applicable when dust explosion risks cannot be fully controlled or mitigated through prescriptive measures as outlined in the standard. This procedure is typically reserved for complex or unique cases where direct regulatory references are difficult to apply, making the Performance-Based Design Option a rarely used approach.

Chapter 7 focuses on "Dust Hazard Analysis (DHA)." The primary goal of DHA is to evaluate and provide management guidance for risks related to fire, flash fire, and explosion associated with dust throughout all stages and environments of the production process. This chapter aims to establish a thorough analysis of these hazards and define reliable prevention and protection methodologies. Responsibilities, applicability, and review frequency are clearly stated in the opening sections.

Chapter 8, "Management System," describes the implementation of procedures and training programs designed to prevent or reduce the catastrophic effects linked to dust presence identified by the DHA. Particular emphasis is placed on effective cleaning of equipment and work areas to minimize dust explosion risks. This chapter also covers the management of Personal Protective Equipment (PPE) regarding selection, usage, inspection, and limitations. Other key topics include incident investigation, procedures for managing organizational or material changes (Management of Change), and the Operational Readiness Review (ORR). The provisions of this chapter apply retroactively.

Chapter 9, "Hazard Management: Mitigation and Prevention," provides detailed prescriptive requirements and methods for preventing and protecting against the consequences of dust combustion and explosions. This chapter is divided into seven sections:

1. **Building Design.** Industrial buildings subject to potential dust combustion or explosion hazards must be designed to prevent and minimize the impact of these events, ensuring the safety of occupants and structural integrity. The section specifies practices related to separation, segregation, and detachment.
2. **Equipment Design.** This section details the design, installation, and maintenance of equipment used in processes involving combustible dust. Topics include dust containment, pneumatic conveying, dust collection, centralized vacuum systems, air-material separators (AMS), air-moving devices, duct systems, and access doors.
3. **Dust Control.** It focuses on controlling fugitive dust emissions, suppressing dust through liquid agents (dust suppression), and using ventilation systems to prevent dust layer accumulation.
4. **Ignition Source Control.** Generally retroactive in nature unless otherwise specified, this section aims to analyze and control potential ignition sources to prevent dust ignition. It covers sources such as hot surfaces, hot work operations, electrical installations, electrostatic discharges, open flames, friction, impact sparks, and lightning.
5. **Pyrophoric Dust.** Reserved for future updates.
6. **Explosion Prevention and Protection.** For equipment with volumes larger than 0.2 m^3 used in dust-handling processes, NFPA 68 and NFPA 69 provide the relevant standards for explosion protection and isolation of devices to prevent dust combustion and explosions.
7. **Fire Protection.** Due to the close relationship between dust explosions and fire, this final section of Chapter 9 addresses the need for manual and/or automatic fire protection measures where fire hazards are identified. Examples include fire extinguishers, sprinkler systems, and fire detection systems.

Chapter 10, "Emergency Planning and Response," is retroactive and represents an addition compared to NFPA 652, which lacked similar guidance. It offers recommendations on structuring emergency plans, maintaining their currency, and sharing relevant information with authorities and stakeholders. Incident scenarios covered include, but are not limited to, fire, flash fire, and explosion.

Below, within Table 5.8, further information on chapters from 21 to 25 of NFPA 660 standard will be provided.

Below, within Table 5.9, further information on the 26 annexes (from Annex A to Annex Z) of the NFPA 660 standard can be found [15].

5.1.3.2 Overview of Regulations and Standards North America: NFPA 68, NFPA 69, NFPA 70, NFPA 77, NFPA 499

In addition to the contents of the NFPA 660 standard, there is a set of NFPA standards that relate, either directly or indirectly, to dust explosions. For practicality, these standards are briefly summarized below. Specifically, they include the following:

TABLE 5.8
Description of Specific Chapters (Chapter 21 to Chapter 25) of NFPA 660

NFPA 660	Title	Brief Description
Chapter 21	Agricultural and Food Processing	Applicable references are provided for the agricultural and food industries concerning the prevention and protection against combustion or explosion hazards related to dust or combustible solid particulate matter.
Chapter 22	Combustible Metals	This chapter focuses specifically on metal powders or alloys that have the potential to ignite or explode, either as a result of manufacturing processes or through their use within these processes. It also outlines the procedure to determine whether a metal powder or alloy should be classified as combustible or non-combustible.
Chapter 23	Sulfur	This chapter addresses the risks related to the handling, processing, or production of sulfur dust or sulfur in molten form.
Chapter 24	Wood Processing and Woodworking Facilities	Wood processing operations can refer to Chapter 24 for the minimum requirements regarding prevention and protection against the effects of combustion and explosion of wood dust.
Chapter 25	Combustible Dusts and Particulate Solids not Otherwise Specified	This chapter applies to dust types not covered in the previous chapters (21, 22, 23, and 24). Due to the broad range of dusts addressed, it modifies and supplements the requirements established in Chapters 1 through 10. The primary goal of Chapter 25 is to establish safe requirements for a wide variety of activities involving dusts that may be prone to combustion or explosion.

TABLE 5.9
List of Annexes in NFPA 660, from Annex A to Annex Z [15]

NFPA 660	Title
Annex A	Explanatory Material
Annex B	Dust Hazard Analysis (Example CMD-FUN)
Annex C	Accumulated Fugitive Dust (CMD-FUN)
Annex D	Supplementary information on Fire Protection (CMD-AGR)
Annex E	Supplementary Information on Fumigation (CMD-AGR)
Annex F	Employee Health and Safety (CMD-AGR)
Annex G	Schematics of Typical Pneumatic Conveying Installation (CMD-AGR)
Annex H	Checklist for Dust Hazards Analysis – Example for an Existing Facility (CMD-AGR)
Annex I	Electrically Conductive Floors (CMD-CMM)
Annex J	Supplementary Information on Magnesium (CMD-CMM)
Annex K	Design for Dust Concentration Control in Ductwork (CMD-CMM)
Annex L	Supplementary Information on Tantalum (CMD-CMM)
Annex M	Supplementary Information on Titanium (CMD-CMM)
Annex N	Supplementary Information on Zirconium (CMD-CMM)
Annex O	Extinguishing Agents That Should Not Be Used on Lithium Fires (CMD-CMM)
Annex P	Testing for Detailed Characterization of Explosive Behavior of Materials (CMD-CMM)
Annex Q	Explosion Protection (CMD-HAP)
Annex R	Informational Primer on Spark Detection and Extinguishing/Diversion System (CMD-HAP)
Annex S	Dust Layer Characterization and Precautions (CMD-HAP)
Annex T	Deflagration Propagation Isolation Methods (CMD-HAP)
Annex U	Use of Water as Extinguishing Agent for Combustible Particulate Solids (CMD-HAP)
Annex V	Guide for Computing Thermal Dose to Personnel from Flash Fires and Building Compartment Deflagrations (CMD-FUN)
Annex W	Conveying System Isolation
Annex X	Automatic Water Spray Deluge Protection for Dryer Systems (CMD-WOO)
Annex Y	Combustible Dust Test Data
Annex Z	Informational References

- NFPA 68:2023 Standard on Explosion Protection by Deflagration Venting.
- NFPA 69:2024 Standard on Explosion Prevention Systems.
- NFPA 70:2023 National Electrical Code.
- NFPA 77:2024 Recommended Practice on Static Electricity.
- NFPA 499:2024 Recommended Practice for the Classification of Combustible Dusts and of Hazardous (Classified) Locations for Electrical Installations in Chemical Process Areas.

The NFPA 68 standard is structured into eleven chapters and thirteen annexes (from Annex A to Annex M). Its primary objective is to safeguard human health and safety, as well as to protect industrial property, through the proper design, installation, and maintenance of explosion venting systems for combustion gases and overpressures resulting from deflagrations. The protective and operational goals set forth in the standard can be achieved using either performance-based provisions or prescriptive-based provisions. The core principles of deflagration venting are outlined in Chapter 6, "Fundamentals of Venting of Deflagrations," while the specifics of venting overpressures caused by gas and mist mixtures, dusts, and hybrid mixtures are addressed in Chapter 7, "Venting of Deflagrations of Gas Mixtures and Mists," and Chapter 8. "Venting of Deflagrations of Dusts and Hybrid Mixtures". For dust explosion scenarios, additional detail is provided in Chapter 9, "Venting of Deflagrations of Gases and Dusts in Pipes and Ducts Operating at or Near Atmospheric Pressure". Further technical insights into deflagration parameter measurements and explosive characteristics of selected dusts can be found in Annex C and Annex F of the standard [16].

The NFPA 69 standard is composed of fifteen chapters and eight annexes (from Annex A to Annex H). Its main purpose, in addition to NFPA 68, is to provide tools and methodologies aimed at preventing an explosion from occurring in the first place. A basic yet important comparison between these two standards highlights a key distinction: while both address explosion-related risks, NFPA 68 focuses on protecting against explosions in confined volumes (enclosures) through the use of venting devices, NFPA 69 emphasizes a preventive approach, as suggested by their respective scopes and content. NFPA 69 is therefore applicable in scenarios where the objective is to prevent the development of explosions in confined volumes that contain combustible substances and materials, such as dusts. The prevention goal can be achieved using one or more of the techniques outlined in the standard, which are detailed in Table 5.10 of the 2024 edition of NFPA 69 [17].

The NFPA 70 standard, commonly known as the National Electrical Code (NEC), is organized into nine chapters and eleven annexes (from Annex A to Annex K). This document holds significant importance as it serves as the primary reference for electrical safety in the United States, particularly concerning the design, installation, and inspection of electrical systems and equipment in industrial facilities and buildings.

TABLE 5.10
Methodologies to Prevent Explosions in NFPA 69 [17]

No.	Description	NFPA 69 Chapter
1	Control of Oxidant Concentration	7
2	Control of Combustible Concentration	8
3	Pre-deflagration Detection and Control of Ignition Sources	9
4	Explosion Suppression	10
5	Active Isolation	11
6	Passive Isolation	12
7	Deflagration Pressure Containment	13
8	Passive Explosion Suppression	14

Given the critical nature of electrical hazards and the rapid pace of advancement in electrical and electronic technologies, NFPA 70 undergoes frequent updates. For example, the 2020 edition was updated in 2023, with further revisions expected by 2026, to stay aligned with technological developments and emerging electrical risks. One of the key connections between NFPA 70 and dust explosions lies in its definition of "combustible dust," which is described as: "solid particles that are 500 μm or smaller [...] that can form an explosive mixture when suspended in air at standard atmospheric pressure and temperature" *(NFPA 70, 70-32, Article 100 – Definitions).* Another crucial link is found in Chapter 5, "Special Occupancies," which addresses the classification of hazardous areas. This chapter outlines the requirements for electrical and electronic equipment based on the classification of these hazardous areas, depending on the type and concentration of combustible materials present. Specifically, NFPA 70 enables the classification of hazardous areas into three classes, each further divided into two divisions, offering a more granular assessment of risk levels related to the presence of flammable gases, vapors, or dust.

The three classes provided for are those reported in Table 5.11 [18].

The NFPA 77 standard is structured into eighteen chapters and nine annexes (from Annex A to Annex I). Its primary purpose is to prevent fires and explosions triggered by electrostatic discharges. The information and recommendations which contains focus on the identification, assessment, and control of electrostatic hazards by managing the generation, accumulation, and dissipation of static charges. The key topics addressed in the standard include fundamentals of static electricity, evaluating static electricity hazards and control of static electricity and its hazards. In relation to dust explosions, Chapter 15 (Powders and Dusts) is particularly relevant. It is dedicated to the interaction between electrostatic charges and powders, examining how this relationship can lead to fire or explosion events. Specifically, the chapter addresses the combustibility of powders and the phenomenon of electrostatic charging caused by contact during powder handling operations. It then examines the concept of powder resistivity, according to which powders are classified as low-resistivity (conductive), medium-resistivity (semi-conductive), or high-resistivity (insulating) materials. Section 15.5 (Discharges in Powder Operations) explores electrostatic discharge events during handling operations. It details the various forms of electrostatic discharges, their distinguishing features, and the specific conditions under which they may occur. The chapter also examines the different methods of powder transport and their impact on static charge accumulation and dissipation. Finally, Chapter 15 also

TABLE 5.11
Hazardous Area Classification in NFPA 70 [18]

Class	Brief Description
I	Location with presence of gases or vapors
II	Location with presence of combustible dusts
III	Location with presence of combustible fibers

provides guidance on hybrid mixtures of combustible dusts with flammable vapors or gases, and the manual dosing of combustible powders into flammable liquids, highlighting the associated risks and control strategies [19].

The NFPA 499 standard is organized into six chapters and two annexes (from Annex A to Annex B). Its primary purpose is to provide guidance for the classification of combustible dusts and hazardous (classified) locations for electrical installations in chemical processing areas and other environments where combustible dusts are produced, handled, or processed. The standard also includes examples that offer practical insight into the reference parameters used to define the degree and extent of hazardous areas. In connection with the provisions outlined in NFPA 70, NFPA 499 serves as a valuable and necessary regulatory tool to guide the selection of electrical systems and equipment intended for safe use in classified areas designated as Class II, Division I, and Division II. The standard provides information on combustible dusts and their possible classifications based on specific characteristics, as well as on combustibility and explosibility testing of dusts, hybrid mixtures, electrostatic charges, and ignition criteria.Finally, the topic of classification of areas with the presence of combustible dust is developed in Chapter 6 (Classification of Combustible Dust Location) [20].

5.2 DUST EXPLOSION RISK MANAGEMENT

5.2.1 Risk Management: An Overview

Throughout history, some industrial accidents have left a lasting mark on collective memory, mainly because of the devastating consequences they caused. The effects on people, the environment, and the economy were often profound. When looking ahead, the most powerful way to avoid repeating such disasters is through prevention [21]. Prevention means putting in place a well-designed risk management plan that follows the principles set out in ISO 31000 (Risk management – Guidelines). This standard goal is not only to safeguard what is valuable but also to create added value by encouraging better performance and innovation [22]. By managing risks proactively, it is possible to reduce the chances that catastrophic events will happen and lessen their impact if they do occur. Moreover, the continuous drive toward improvement and innovation naturally follows from embracing the Plan-Do-Check-Act (PDCA) cycle, a proven approach for keeping risk management plans effective and up to date. The connection between the PDCA framework and ISO 31000 structure is clearly outlined in Table 5.12 [23].

However, it is important to emphasize that even the safest facility can never be considered completely risk-free. For this reason, it becomes essential to deepen the risk analysis to a level that is technically and economically acceptable, evaluating possible safety implementations in terms of tools, procedures, and response measures. The impossibility of zero risk stems from the fact that risk is often inherent in the very nature of work operations, the characteristics of the process, materials, and equipment used. Therefore, the purpose of risk assessment is first to identify these risks, analyze them, and define an effective control framework that must be

TABLE 5.12
Relationship between PDCA and the Structure of ISO 31000:2009 [23]

PDCA Element	Chapter in ISO 31000:2009	
Plan	Design (of risk management structure)	5.4
Do	Implementation	5.5
Check	Monitoring and Review	6.6
Act	Continual Improvement	5.7

periodically updated [21]. The adoption of a risk management plan is crucial for all types of companies, with particular attention to small and medium-sized enterprises (SMEs), which often, due to their structure and available resources, are more vulnerable compared to larger companies. Empirical evidence of this is that around 50% of SMEs close before reaching their fifth year of operation. Considering that the global industrial fabric is mainly made up of small- and medium-sized companies, a broader implementation of risk management plans could be a key factor in promoting sustainable economic and social development worldwide [23].

In general, therefore, the risk management process provides a structured method to understand the risks associated with industrial activities and to achieve acceptable risk levels by implementing appropriate prevention and/or mitigation measures that are specific to each company [21]. Furthermore, a functional risk management plan must be able to integrate risk analysis within the company's core activities and functions, directly involving both governance and management [22] and the very structure of the risk management plan must be clearly defined from the outset, with well-established objectives to be achieved [21]. A good starting point for developing a risk management plan could be to properly evaluate existing risk analysis procedures, assess their applicability to work processes, identify any gaps, and address those shortcomings within the analysis process [22].

The Risk Management plan is mainly structured into three sequential steps described in the following and illustrated in Figure 5.5.

- **Risk Assessment**. The methodological approach of risk assessment is fundamentally linked to the need to evaluate and reduce the level of a risk until it reaches a tolerable threshold. The process is divided into three sub-steps:
  (a) Risk Identification. Identification of the hazard and characterization of the associated risks.
  (b) Risk Analysis. Analysis of the severity of consequences and the frequency of occurrence associated with a single risk.
  (c) Risk Evaluation. Evaluation of the acceptability of the risk based on pre-established analysis tables [22].
- **Risk Control (or Risk Treatment)**. Definition of prevention and protection measures to mitigate the risk with approaches that may fall within different

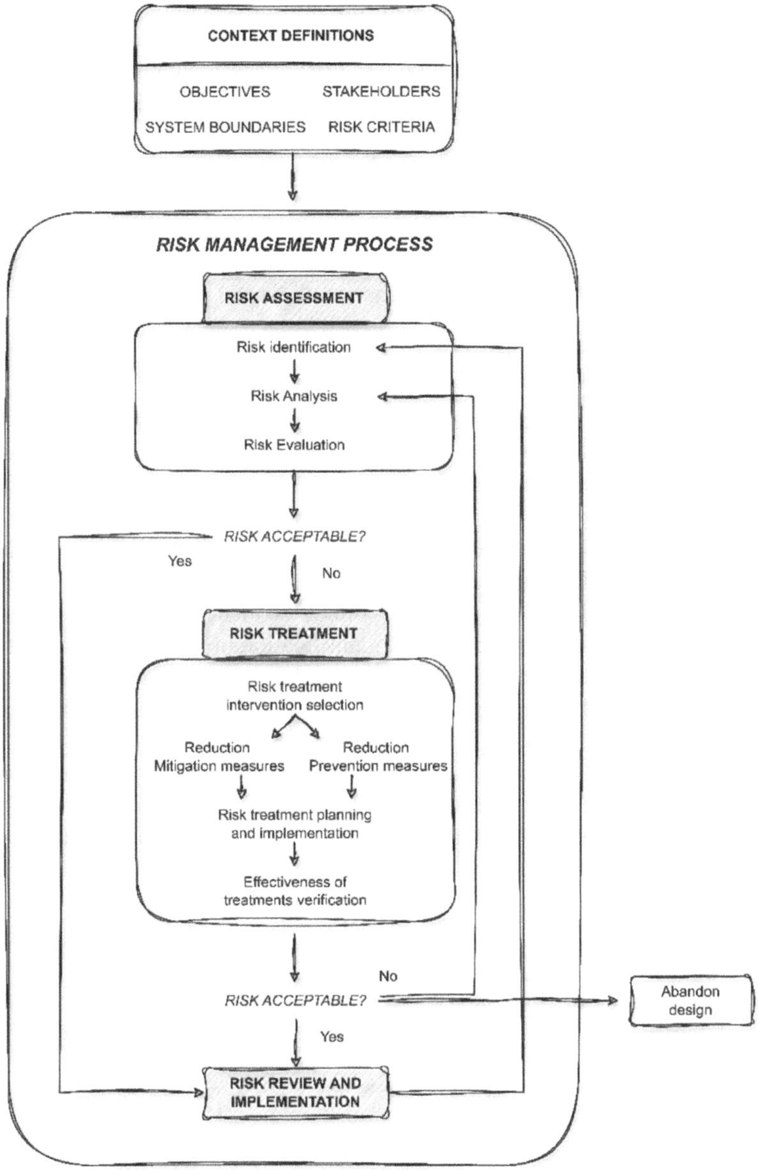

FIGURE 5.5 Risk management overview [21].

levels, as shown in Figure 5.7, with the aim of reducing the frequency of occurrence and the severity of consequences. Once the reduction measures are defined, it is necessary to repeat the risk assessment process to reassess the extent of the scenario. When this is considered acceptable, it is possible to proceed.
- **Risk Review**. Through auditing and monitoring activities, it is possible to implement and readapt the measures adopted for mitigating the incidental event in case the identified risk undergoes any variation or modification (see Figure 5.5). Supporting supervision through audits and monitoring can also include updated training courses, case studies, and post-incident investigation results [21].

5.2.1.1 Risk Assessment

As the first building block of risk management, risk assessment is the process through which it is possible to estimate the risk associated with various events or states, allowing the identification of opportunities or needs for structural or procedural changes. The purpose of risk assessment is, therefore, to describe the origin, probability, and magnitude of the risk associated with a substance, equipment, or scenario where a significant degree of uncertainty is present [24]. The assessment process can thus be described as an organized sequence of activities, which, as shown in Figure 5.5, begins with the risk identification process. The aim of this initial step is to support the finding, recognition, and description of the risk in terms of origin, probability, and magnitude. A representation of the risk identification process is shown in Figure 5.6 [22].

The importance of the risk identification process can be summarized by a phrase from Crispin G.: "unidentified risks cannot be mitigated." For this reason, a significant portion of the risk management process should be dedicated to risk

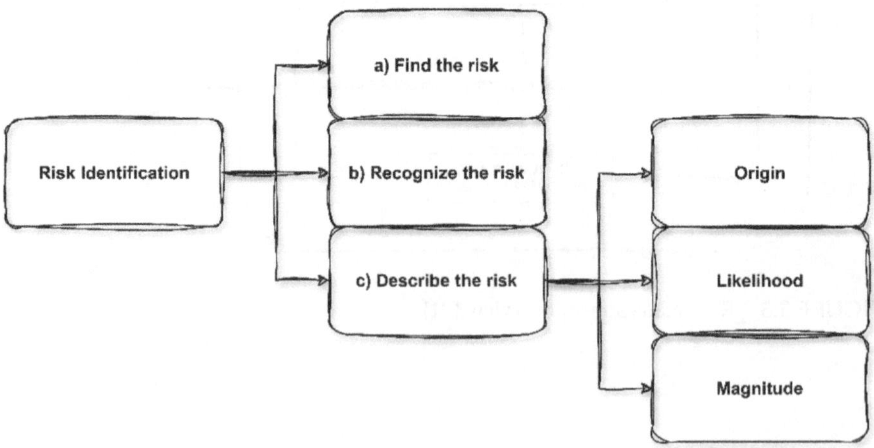

FIGURE 5.6 Risk identification process.

identification, since a proper identification allows for understanding the nature of the risk and how it can be effectively managed. To identify risks, it is useful to begin the activity with a brainstorming session to define the analysis context, using specific techniques and tools to gather as much information as possible to identify the greatest number of risks present [25]. The second fundamental step of risk assessment is risk analysis, which can be defined as the process through which the identified risk is analyzed at various levels of detail depending on the availability of reliable information and the purpose of the analysis. A crucial aspect of this process is the influence of factors such as biases and risk perception on the evaluation of the risk level. For this reason, it is important to consider these aspects in relation to the final analysis before proceeding to the Risk Evaluation step [22].The third and final step is Risk Evaluation, which is the phase where the acceptability of the risk identified through Risk Analysis is assessed using predetermined acceptability tables [22]. At this point in the Risk Assessment procedure, it is essential to ask: "Is the risk acceptable?" It is also important to define what is meant by acceptable risk, which can be defined as:

> *risk whose probability of occurrence is so small or whose consequences are so slight or whose benefits (perceived or real) are so great that individuals or groups in society are willing to take or be subjected to the risk that the event might occur. An acceptable risk requires no risk management; it is, by definition, acceptable. A risk that is not acceptable is, therefore, unacceptable and by definition must be managed* [24].

However, despite the risk having been identified and analyzed and the definition of acceptable risk being established, determining whether a risk is truly acceptable is not a straightforward scientific task because many aspects must be considered, including the individual perspectives of various stakeholders. The reference standard for the techniques and tools that can be used for the entire risk assessment process, as described in ISO 31000, is IEC 31010 [26]. Among the techniques cited in Annex B of IEC 31010 are those reported in Table 5.13.

5.2.1.2 Risk Control or Risk Treatment

In reference to what is reported within the ISO 31000 standard, it is possible to intervene on the risk identified through the Risk Assessment procedure by means

TABLE 5.13
Risk Assessment Techniques [26]

Activity	Techniques
Eliciting views from stakeholders and experts	Brainstorming
Risk identification	Checklist, FMEA, FMECA, HazOp, ROA
Analyzing controls	Bow Tie, HACCP, LOPA
Consequences and Likelihood Understanding	BIA, ETA, FTA, HRA
Selecting between options	CBA, Game Theory, Multi-Criteria Analysis

of various targeted actions aimed at achieving an acceptable level of risk [22]. It is also important to remember that the acceptability of risk cannot have an absolute value but must be defined considering the context within which the analysis activity takes place [27]. The process through which risk acceptability can be reached is an iterative procedure that takes into account a series of sequential steps arranged in a logical order. Specifically, as reported within the ISO 31000 standard, these steps are: (a) formulation and selection of various risk treatment interventions; (b) planning and implementation of risk treatment; (c) verification of the effectiveness of the treatment; (d) evaluation of the residual risk level and determination of whether it is acceptable or unacceptable and, in this last case, further treatment interventions must be planned.

After formulating various hypotheses about which treatment interventions might be suitable for the specific risk under analysis, it is necessary to proceed with selecting the best proposal among those available. The selection activity of the interventions, in this case, is defined within the ISO 31000 standard as a process of identifying the intervention or interventions that offer the best balance between costs and benefits directly linked to the planned treatment, also taking into account the opinions of the various stakeholders involved. The approaches that can be considered for risk control or treatment may aim to modify the frequency or magnitude of the risk itself or, when possible, eliminate the source of risk or the activity [22]. A useful diagram related to this topic is reported in Figure 5.7, which provides a progression of the

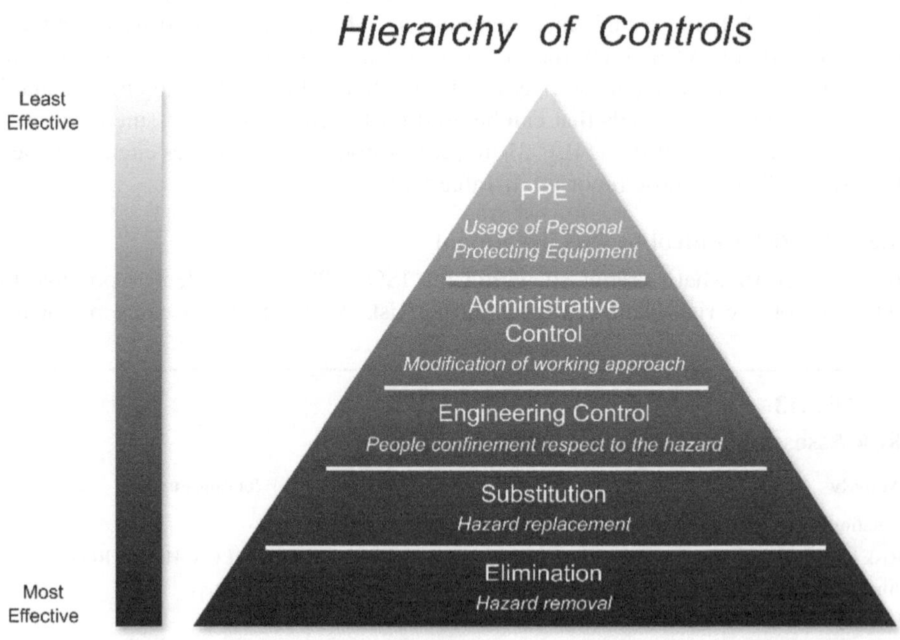

FIGURE 5.7 Hierarchy of risk controls.

effectiveness of the various macro-interventions for risk treatment, starting from the most effective to the least effective.

The planning activity for risk treatment involves the structured organization of the treatment implementation process in such a way that it is clear how actions will be carried out and possible to monitor the progress of the plan. Among the information that must be included in the treatment plan are the reasons for choosing specific interventions and the expected benefits, the distribution of responsibilities, the necessary resources, and the expected time frame for completing the planned activities [22].

Concluding, it should be noted that ISO 31000 (as an international reference standard on Risk Management, 2018 version) does not provide in-depth details regarding the above-mentioned activities of verifying the effectiveness of the treatment and evaluating the acceptability of the risk.

Given the complexity of the topic of risk acceptability, it is important to highlight that this is the result of a cognitive judgment process which, in the case of risk analysis, is linked to an evaluation policy based on scientific evidence [28]. Therefore, once the process of evaluating the reliability of the treatment has been defined, the next question is whether the residual risk is sufficiently low to be accepted as is. In other words, the question can be phrased as: "how low is low enough?" [29]. In fact, as can be inferred, the process of determining risk acceptability also cannot have absolute value but it is subjective and dependent on the context of the analysis [28].

5.2.1.3 Risk Review

As the third and final step among those listed in section 5.2.1, it is essential to adopt a Risk Review practice, whose objective is to oversee the entire Risk Management process and implement potential improvement measures as part of a periodic review of the overall plan. The aim is to update and adapt Risk Management activities to reflect the natural developments and changes that, over time, may affect facilities and process operations, as well as to identify any new risks that need to be analyzed and addressed. In its entirety, Risk Management (with all its components) is a process that must be documented, recorded, updated, and properly maintained so that it remains accessible and available for information-sharing activities with external entities, for conducting critical evaluations, and for making business-relevant managerial decisions [22].

5.2.2 Dust Hazard Analysis (DHA)

With the aim of providing a more convergent view of the tools cited in the previous section 5.2.1 as they relate to dust explosions, this section will delve into the topic of Dust Hazard Analysis (DHA) in order to facilitate understanding of a type of analysis that can be conducted to improve workplace safety and protect human life, the environment, and industrial assets against dust explosion hazards. Dust Hazard Analysis (DHA) is defined as a systematic, documented review of processes and work areas where combustible dust is present, in order to identify potential risks of fire, flash fire, or explosion. The objective of this process is therefore to produce recommendations that allow for the appropriate management of identified risks [15], through the identification of hazards that, in some cases, may not have even been previously known [30]. Within the NFPA 660 standard, Chapter 7 provides regulatory references for DHA and outlines its scope and limitations [15]. Applying the DHA method enables the classification of the analyzed areas into three categories based on the level of identified hazard. These areas are:

1. No hazard. Once the DHA procedure identifies the analyzed area as free from hazards, it is possible to avoid further investigations or the installation of additional prevention measures to improve the area's safety level.
2. Potential hazard. This category represents an intermediate case in which further investigation is required through additional methods, testing, and tools in order to clearly and comprehensively determine whether the analyzed area falls into a hazard or no-hazard classification. The existence of this intermediate category is intended to ensure heightened attention to risks arising from the presence of combustible dusts, as it mandates a deeper level of investigation whenever an area is classified as potentially hazardous, thus ensuring a more reliable analysis.
3. Hazard.
   a. Fire. The identification of a fire scenario through DHA is tied to the recognition of at least one potential ignition and combustion event involving dust. This, in turn, can lead to the involvement of other materials and substances that may sustain the fire within the work area.
   b. Flash fire. A second type of accident scenario that can be identified through DHA is flash fire, which is defined in the NFPA 660 standard as "a deflagration that spreads rapidly through a flame front in a dispersed fuel such as dust, without the production of damaging overpressure waves." In any case, identifying this type of scenario allows for the implementation of suitable prevention and protection measures to safeguard personnel health and safety.
   c. Explosion. This is the most severe scenario that may be identified through the DHA method. The risk of dust explosion can lead to very serious consequences due to the destructive impact of overpressure waves on people, structures, and equipment, which, in turn, can trigger disastrous domino effects. Applying a DHA can therefore enable the recognition of such events and provide the opportunity to implement appropriate prevention and protection tools against dust explosions [31].

The application of a reliable and case-specific risk assessment procedure represents an important means of prevention. For this reason, a well-structured DHA has the potential to prevent the development of the incident scenarios described above. However, it is important that this investigative tool is organized in a clear, simple, and precise way so that it can be easily applied and understood by all concerned personnel. In this regard, several steps can be recommended to characterize an effective and comprehensive DHA.

1. **Identify the presence of combustible dust in the work process.** The characterization of the dust can be carried out through the use of laboratory tools capable of analyzing specific properties related to combustibility and explosiveness, in order to determine whether it can be classified as combustible.
2. **Define which standard must be applied.** Depending on the geographical location of the facility, different applicable standards may need to be considered. These may include the standards referenced in section 5.1.
3. **Organize and create the work team.** As suggested in the NFPA 660 standard, the DHA must be conducted by qualified personnel with expertise in dust explosion hazards.
4. **Collect the necessary documentation.** At this stage, it is important to gather all relevant documents, such as: plant diagrams, work and cleaning procedures, emergency plans, safety reports, inspection records, and more.
5. **Obtain useful data to identify fire, flash fire, or explosion scenarios.** The main goal of the DHA is to identify likely incident scenarios that require protection measures. The analysis should allow an evaluation of the incident scenario and potential ignition sources, considering both magnitude and probability, in order to determine whether the scenario is likely to occur.
6. **Evaluate safeguards in relation to the applied standards.** Given that safeguards may already be present in the facility analyzed with the DHA, and considering the likely incident scenarios identified in point 5, it is necessary to evaluate whether these safeguards are adequate and compliant with the standards identified in point 2.
7. **Assess the need to integrate existing safeguards with additional ones.** Based on the evaluation of the existing safeguards, a decision can be made regarding the need for further implementation to ensure an adequate level of protection against likely incident scenarios.
8. **Verify the impact of these integrations on the identified scenario consequences and probabilities.** The implementation of safeguards may influence the probability and magnitude estimation of the reference incident scenario. Therefore, it is necessary to re-evaluate both probability and magnitude in light of any new implementations.
9. **Document the entire DHA process and recommendations in accessible and consultable formats.** In order for the DHA procedure to be available when needed (e.g., during inspections, updates, or other uses), the entire analysis and implementation process must be documented and stored appropriately so that it is easily retrievable.
10. **Implementation of recommendations.** If necessary, the recommendations should be implemented correctly and within the expected timeframes so they can effectively contribute to the prevention and protection against the incident scenarios identified through the DHA.

The approach used in the DHA is similar to that provided for the application of the PHA (Process Hazards Analysis), which consists of an organized and systematic method for identifying and analyzing risks primarily associated with the presence of chemical substances in industrial processes. The key difference lies in the focus, which shifts from chemical substances to combustible dusts [31].

Summarizing, among DHA scopes there are: (a) individuate eventual connections between instruments or building compartments conditions that can facilitate the diffusion of explosion through the plants; (b) define safety operating parameters that, such as equipment temperature and pressure or maximum level of dust accumulation in the working area; (c) identify existing safeguards applied on equipment or buildings to manage properly fire and explosion hazards; (d) if necessary, propose integrative strategy, measures and safeguards [32].

5.3 CHARACTERISTICS OF DUST EXPLOSIVE ATMOSPHERE ENVIRONMENTS

5.3.1 Comparison of Combustion Characteristics of Dust and Gas/Vapor

From a technical standpoint, it is possible to state that explosive dust clouds and explosive gas/vapor clouds, once formed, share similar characteristics in terms of ignition and combustion behavior. However, it is important to emphasize that these two types of explosive clouds each have specific features that are not applicable to the other. Therefore, despite the aforementioned similarities, there are significant differences between dust cloud explosions and gas/vapor cloud explosions, which must be considered starting from the formation phase of the clouds themselves [33]. The most relevant differences are listed in the following, and they will be briefly described for clarity: (a) necessary conditions for a deflagration; (b) chemical purity of fuel; (c) particle size and shape; (d) uniformity of fuel concentration and initial turbulence; (e) range of ignitable fuel concentrations; (f) heterogeneous and homogeneous chemical reactions; (g) incomplete combustion.

(a) Risk analyses and investigative procedures often take into account the different conditions required to trigger a gas/vapor explosion versus a dust explosion. In the first case, the simultaneous presence of the three elements of the fire triangle (fuel, oxidizer, and ignition source) is sufficient. In contrast, dust explosions require two additional elements, forming what is known as the explosion pentagon. These additional elements are suspension of the dust (which can lead to a flash fire) and confinement (which can result in a dust explosion). Another important aspect is smoldering, a slow combustion process that can propagate within layers or accumulations of dust. If stimulated with sufficient intensity, it may lead to the dispersion and ignition of a dust cloud.

(b) The chemical composition of dust can be subject to alterations that must not be underestimated when evaluating its explosive potential. Unlike gases or vapors, which can be readily characterized in terms of chemical composition even when dispersed in cloud form, dust has a high potential to adsorb external substances (especially air and water). These adsorbed substances act as contaminants capable of altering the chemical-physical and explosive properties of the "pure" dust (particularly the outer surfaces of dust layers and accumulations) [34].

(c) Unlike gases and vapors, the explosive characteristics of a dust are also influenced by the size of its particles and their shape, as indicated in Section 3.1 of this book.

(d) The formation dynamics of a flammable gas and vapor cloud differ significantly from those of a dust cloud. In the case of gases/vapors, the formation of a cloud with ignition-relevant concentrations is time-dependent: the longer the emission source releases gas and vapors, the higher their concentration within the cloud. Once the Lower Flammable Limit (LFL) is reached, the cloud can ignite. The formation of a dust cloud, on the other hand, follows a different process because it requires a physical stimulus (e.g., a primary explosion, air movement, or other events) to lift a dust layer or accumulation into suspension and create a cloud. Moreover, the development of a dust cloud rarely follows a pattern of homogeneous dust concentration distribution; instead, the cloud is characterized by uneven concentrations. The formation of a dust cloud is also influenced by a time-related effect opposite to that of gases and vapors: while the passage of time increases the concentration in the latter, in the case of dust, particles are increasingly subjected to settling. This settling process causes the cloud, if no ignition occurs, to revert to a layer or accumulation form [34].

(e) For a cloud of substance to be considered combustible, the concentration of the substance must fall within the flammability range specific to that substance. For air-gas or air-vapor mixtures, these concentration limits are expressed as LFL (Lower Flammable Limit), which is the minimum concentration necessary for ignition below which combustion does not occur due to lack of fuel, and UFL (Upper Flammable Limit), which is the maximum concentration above which combustion does not occur due to lack of oxidizer [35]. In contrast, the flammability range of dusts is defined by the MEC (Minimum Explosive Concentration), which is similar to the LFL, especially for fine powders and represents the lower limit. For most dusts, no specific upper limit is reported above, which the cloud cannot ignite due to fuel excess [36].

(f) Homogeneous reactions involve both reactants and products in the gaseous state, while heterogeneous reactions occur in systems where two or more phases (solid, liquid, gas) are involved in various combinations. Based on this, combustion, as a chemical oxidation process, can be classified as either homogeneous or heterogeneous [37]. The combustion reaction of a flammable gas or vapor occurs in the gas phase, except for minor contributions, and can, therefore, be defined as a homogeneous combustion reaction. The combustion reaction of dust, instead, involves a complex system with both homogeneous and heterogeneous reactions [34]. The presence of a solid fuel, like dust, influences the type of combustion reaction and leads to different development compared to homogeneous combustion. For example, the combustion of organic dusts found in many food and pharmaceutical industry processes involves a complex interaction of both reaction types. In a cloud of solid particles, heating causes pyrolysis of the material, releasing combustible vapors into the air. When these vapors mix with air in concentrations within flammability limits, flame propagation may occur through homogeneous gas-phase reactions. Pyrolysis leaves behind a carbonaceous solid residue that can oxidize through heterogeneous reactions at the solid-gas interface. The hazard of dust fires also lies in the overlap of these reactions, where gaseous combustion and solid surface combustion coexist and influence each other, with heat from gaseous combustion accelerating solid combustion and vice versa [38].

(g) A final aspect that distinguishes the combustion of dusts from that of flammable gases or vapors is the completeness level of the combustion process itself. In the case of flammable gases or vapors, complete combustion is generally expected, whereas dust combustion tends to be incomplete. Complete combustion is defined as a reaction in which all oxidizable elements (mainly carbon (C), hydrogen (H), and sulfur (S)) are converted into CO_2, H_2O, and SO_2 based on stoichiometric reaction proportions and the availability of oxygen as the oxidizer. Incomplete combustion, instead, occurs when oxidation takes place with insufficient oxygen relative to the stoichiometric proportions, resulting in only partial transformation of the combustible substances into the products mentioned above. To achieve complete combustion, an excess of oxygen relative to the stoichiometric fuel requirements is necessary [39].The theoretical air volume would be sufficient for complete combustion only if there were a thorough mixing of fuel molecules, which can occur only under theoretical conditions. To achieve this level of combustion, it is, therefore, necessary to operate with excess air because, in real conditions, it is not possible to attain the

TABLE 5.14
Percentages of Excess Air for Complete Combustion Based on the Physical State of the Fuel [8]

State of Combustible	Excess of Air (%)
Gas/vapor	5–20
Liquid	15–30
Solid	30–100

optimal mixing conditions, which become increasingly complex as the degree of matter aggregation increases. Therefore, it is possible to estimate achieving complete fuel combustion with the excess air percentages reported in Table 5.14 [8].

5.3.2 Dust Explosive Atmosphere Environments: Zone Characterization

In relation to what was outlined in Section 5.3.1, it is evident that the formation of ATEX zones due to the presence of dust has different characteristics compared to those caused by flammable gases or vapors. This has been a valid reason for developing distinct and specific reference standards for the two cases, in order to allow for more targeted interventions depending on the situation in question. Focusing primarily on the reference standards applied within the European context, an attempt will be made below to outline a framework for characterizing ATEX zones resulting from the presence of combustible dusts. For this section, the following standards, already mentioned in Section 5.1, will be used as references: CEI EN 60079-10-2:2015 and ATEX Directive 153.

5.4 HUMAN BEING IN RISK ANALYSIS

The prevention system for incident scenarios is tasked with analyzing all potential causes that could lead to disastrous events such as dust explosions. Among these causes, it is crucial to consider the human factor, often underestimated, yet recognized as the root of a significant portion of incidents, both occupational and process-related, accounting on average for 70–90% of all cases [40]. Given the critical role that operators play within a working process, their involvement should be given prominent attention in risk assessments aimed at reducing and mitigating incidents, especially those related to dust explosions. In this context, the human being must be considered not only as an individual, but also as a component of the human–machine system, and as a participant in interpersonal dynamics. What emerges from this brief introduction is a multifaceted and evolving picture, one that makes it impossible to treat the human element as a static variable when it comes to occupational and process safety. It is, therefore, reasonable to assume that any risk analysis that aspires to be even minimally effective must take into account this inherent complexity and dynamism [41].

5.4.1 Human Factors and Human Reliability Analysis (HRA)

The workplace and everything that defines it will never be perfect. While this may seem like a trivial observation, it is actually crucial to understanding the significance of the human role in building a meaningful safety culture. In an environment where imperfection is inherent, it is the human being who must step in, bridging the system gaps and upholding those procedures and techniques that have already proven effective in preventing harm and protecting health and the environment. The following statement captures this idea with clarity: the presence of people in the workplace is not just inevitable, but a fundamental resource (even, and especially, from a safety perspective):

> *does not explain why human performance practically always goes right. The reason that things go right is not people behave as they are told to, but that people can adjust their work so that it matches the conditions. As systems continue to develop, these adjustments become increasingly important for successful performance....Safety management should therefore move away from ensuring that 'as few things as possible go wrong' to ensuring that 'as many things as possible go right...*
>
> **(McLeod R.W., 2015)**

In this context, it becomes particularly useful to define what is meant by the term *Human Factor*, also referred to as *Ergonomics* or *Human Factors Engineering (HFE)*. Broadly speaking, it is the study of how humans interact with the various elements of their working environment, including tools, systems, technologies, and other people. The primary aim of this discipline is to support safety professionals in identifying the underlying causes of human error within organizational processes and, as a result, help prevent potential accident scenarios [42]. The core approach of Human Factors is to focus on the operator's ability to correctly carry out their tasks, taking into account both human limitations and capabilities. This includes not only routine operational tasks but also non-routine actions, such as those required during emergencies involving fires or explosions [43]. In essence, Human Factors is concerned with key aspects that directly influence workplace activities and the safety with which they are performed. These include: the purpose of the work being carried out, the person responsible for performing it (considered in light of their individual abilities) and the work environment itself, understood not just in physical terms but also as an industrial–organizational context. All of this is further shaped by a broader social framework, both locally and nationally.

TABLE 5.15
Characterizing Factors of Human Performance According to Fitts [44]

Factor	Description
Learning	The aim of the training is to provide professionalism and skills.
Individual differences	Subjective characteristics that justify personalized selection and training plans and that allow for the availability of resources with different potential for different job tasks.
Motivation	Stimulus on which the company must work to incentivize the operator's performance.
Environment	Work context that influences work activities.
Task	Work objective to be achieved which, under certain conditions and for efficiency reasons, can be reviewed periodically.

In detail, the capabilities and performances of an individual, as reported by Paul Fitts several years ago, are a function of five distinct factors, reported and briefly described in Table 5.15.

The scope of Human Factors, however, is primarily limited to improving working conditions through the implementation of new designs aimed at optimizing human performance in relation to productivity and safety. To complement this approach, it is possible to employ Human Reliability Analysis (HRA), which, unlike the Human Factors approach, focuses on predicting human performance and is, therefore, firmly positioned within the field of reliability engineering. A HRA practical application example is presented in Chapter 7. These two approaches, while differing in perspective, can work synergistically to improve the quality of human activities, each compensating for the limitations of the other and ultimately forming a more complete tool than either one could offer individually. In particular, HRA can enhance the Human Factors approach through the following contributions: (a) support in modeling human performance; (b) support in translating Human Factors into an engineering framework; and (c) support in applying predictive modeling of human performance to develop designs that better align with human capabilities in the workplace [12].

It is, therefore, essential to define the concept of "human reliability," which in engineering terms (based on its close relationship with HRA) is expressed as the probability that a specific task will be successfully completed by an individual [45]. Efforts aimed at monitoring and predicting the probability that a task is carried out successfully are ultimately intended to prevent what is known as "Loss of Human Reliability," a situation in which operators perform in ways that are inconsistent with organizational expectations while executing their assigned tasks. From this perspective, another key concept can be introduced: "Human Error," defined as an event in which a specific task is either not carried out or a prohibited action is performed, with potential consequences ranging from minor to catastrophic (such as production delays or even the loss of human life) depending on the nature of the triggered event.

A human error can occur for a variety of reasons, such as making the wrong decision in response to a problem, lack of time, failure to perform an assigned task, performing a task that was not assigned, and inability to recognize a hazardous situation [46].

In addition to these aspects, human error may also stem from forgetfulness, distraction, or a failure to acknowledge that a message was not properly understood, all of which are particularly prevalent in high-stress environments or under demanding circumstances. Given that it is virtually impossible to reach a condition where human error can be entirely eliminated, it becomes essential to manage both organizational structures and work activities through a system capable of analyzing the human element in the greatest possible detail. This analysis should be integrated with all other risk assessment tools which, traditionally, tend to focus more on system-related risks rather than those arising from operator involvement.

HRA is, therefore, a risk analysis tool that considers likely incident scenarios to investigate the causes of human errors and the potential consequences they may trigger, with the ultimate goal of providing a probabilistic forecast of the overall safety of

TABLE 5.16
Relationship between Probability and Type of Human Error [43]

Probability	Description of Error Type
1.0 – 0.1	Processes involving creative thinking, unfamiliar operations, a short period of time, or high stress.
0.1 – 0.01	Errors of omission, in which the dependence is on situational cues and personal memory.
0.01 – 0.001	Commissioning errors, such as using the wrong buttons, reading the wrong display/indicator, etc.
0.001 – 0.0001	Mistakes regularly made in common tasks.
0.0001 – 0.00001	Extraordinary errors that are difficult to imagine how they could occur, occurring in stress-free environments with powerful signals.

the work system [47]. As previously mentioned, human error can occur at any stage of the work system, and the probability of such occurrences can be estimated based on the values reported in Table 5.16. This estimation does not take into account the potential influence that drug or alcohol use may have on human reliability:

There are various HRA methodologies that can be applied to enhance workplace safety. In line with the purpose of this book, a detailed explanation or application of these methods will be avoided; however, a useful list will be provided as a reference for those interested in pursuing further study or considering their application specifically within the food and pharmaceutical industries: PC (Paired Comparisons), APJ (Absolute Probability Judgment), SHARP (Systematic Human Action Reliability Procedure), HEART (Human Error Assessment and Reduction Technique), COGENT (Cognitive Event Tree), and CAHR (Connectionism Assessment of Human Reliability) [47].

5.4.2 Risk Communication

Within the scope of this book, the concept of "risk communication" can be understood as the transmission of essential information necessary to make informed judgments and take autonomous, conscious actions regarding risks that affect health, safety, and the environment. Therefore, risk communication is a fundamental step in ensuring enhanced safety at every level within a company. However, the complexity of the organizational and social systems we live in, combined with the rapid pace of action and reaction these systems demand, sometimes does not allow enough time to properly convey or process information that, due to its importance, could be crucial in preventing an incident within a workplace. Providing clear and targeted information, and being able to correctly receive and contextualize it in the work environment, can enable appropriate and optimal decision-making to protect both the health and safety of people at work and the surrounding environment. It is thus necessary to equip ourselves with tools that allow for "successful communication," enabling individuals to recognize when a particular risk is significant enough to deserve substantial attention and, when risks are shared, to foster the exchange of information among those involved. Nevertheless, to date, no objective method has been identified that guarantees a message has been clearly expressed and fully understood by the recipient [48]. Focusing risk communication on risk prevention means considering it as an exchange of information and concepts within the company and with external stakeholders involved in prevention activities. Many categories of people are involved in workplace safety processes, and effective preventive and protective measures cannot exist without adequate communication. If communication is lacking, the outcome is likely to be unsatisfactory and confusing [38]. To better characterize what "successful communication" entails, it is important to highlight the main objectives of risk communication.

1. Promote awareness and understanding of the specific issues under examination during the risk analysis process among all participants.
2. Foster consistency and transparency in the development and implementation of risk management decisions.
3. Provide a solid foundation for understanding proposed or implemented risk management decisions.
4. Improve the overall effectiveness and efficiency of the risk analysis process.
5. Contribute to the development and delivery of effective information and training programs, when chosen as risk management options.
6. Promote public confidence in food supply safety.
7. Strengthen working relationships and mutual respect among all participants.
8. Encourage appropriate involvement of all interested parties in the risk communication process.
9. Exchange information about knowledge, attitudes, values, practices, and perceptions of stakeholders regarding food-related risks and related topics.

Risk communication must include not only interpersonal exchanges but also the transmission of messages through signs, symbols, and documents. In this regard, it is interesting to mention an analysis by the Chemical Safety and Hazard Investigation Board (CSB) on 281 dust explosion incidents occurring between 1980 and 2005. This analysis revealed that, often in contexts where devastating explosions took place, workers and managers were unaware of the potential dust explosion risks or failed to recognize the severity of this hazard. Among these 281 incidents, the CSB found that 140 safety data sheets for products capable of producing combustible dust were either missing or lacked adequate communication about this hazard. Specifically, 41% of the sheets contained no information about explosion risks, and 59% provided information that was difficult to understand [49].

Such communication gaps and the questionable availability of information about potential dust explosion risks must be carefully considered during risk analysis. Operators who, in various ways, come into contact with dust during work processes should receive adequate basic information from their employers about the risk of dust explosions. This process, an essential part of mitigating dust explosion risk, relies on appropriate communication methods. Basic communication could be delivered, for example, through a brief informational letter, supported by awareness videos, covering the following points: (a) the fact that dust explosion hazards exist; (b) basic chemical and physical concepts of dust explosions and their triggers; (c) how to extinguish a fire; (d) how to perform proper cleaning and information on secondary explosions; (e) why it is important to report incidents and damages.

Moreover, understanding the fundamentals of risk perception is crucial for structuring effective risk communication [50].

5.4.3 Risk Perception

Risk perception rightfully belongs among the key elements to consider during a Risk Management procedure, since Risk Management aims to reduce both the likelihood and severity of disastrous events through the development of prevention and protection strategies. The crucial connection between Risk Perception and Risk Management lies in the fact that risk management actions are influenced by how people perceive risk, and conversely, a strong focus on preparing for a specific risk can lead to that risk being perceived as less significant. In essence, this dynamic relationship can be illustrated by the image in Figure 5.8 [51].

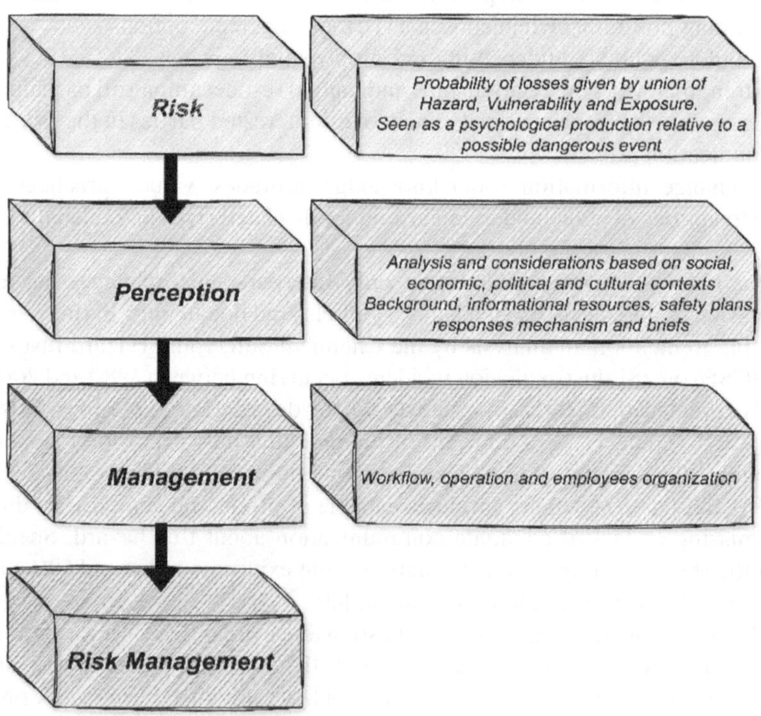

FIGURE 5.8 Evolutive connection between risk management and risk perception [51].

Perception can be described as a mental process of interpreting and assigning meaning to various stimuli, shaped by each person's emotional conditioning, experiences, and imagination. When this process is applied within a scenario where the stimulus aims to communicate a risk, it evolves into an individual evaluation of the likelihood and potential consequences that a certain event might cause. It is also important to consider how people often base their risk judgment on information that already supports their existing view of the risk or problem, reinforcing their beliefs regardless of the actual facts or scientific data. Additionally, there is a common tendency to be influenced by information that is easy to remember and to perceive situations as more dangerous when there is less information available. These are just a few findings from various studies that conclude risk perception differs from objective facts, a conclusion supported by over 126 psychological models and theories developed over the years regarding risk perception.

Nonetheless, although risk perception is personal and varies from person to person, a familiar risk tends to cause less harm and milder consequences compared to a less familiar one. The factors that most influence risk perception include: familiarity

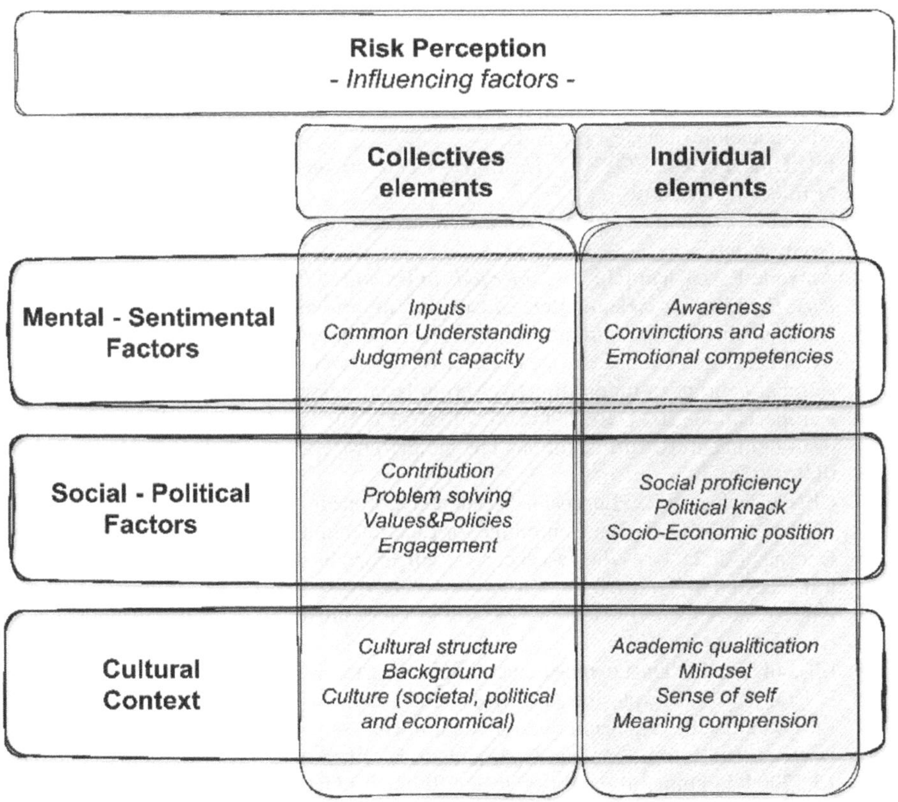

FIGURE 5.9 Elements that influence the subjective perception of risk [51].

with the risk source or past experiences, voluntary or involuntary exposure, ability to control the risk and, personal benefit from the risk source.

However, by considerably broadening the perspective to include a wider range of factors that can influence subjective risk perception, it is possible to categorize these elements as outlined in Figure 5.9.

The importance that should be given to risk perception within the framework of workplace safety can be summarized by a quote attributed both to the social psychologist W.I. Thomas and to the sociologist Émile Durkheim. The quote, in any case, states: *"Things which are perceived as real will be real in their consequences."*

A direct application of this statement to the concept of risk lies in the fact that risk itself can be considered an abstract notion, carrying different meanings for different people: for a safety professional, it may represent the possibility of an accident; for a manager, it could mean a greater or lesser probability of business success; while for a financial agent, it might indicate the jeopardizing of investments depending on stock market trends. Risk, therefore, is imbued with subjective value, and its perception shapes a reality that must be incorporated into every risk analysis and management activity to enhance the effectiveness of prevention and protection tools against industrial accidents [29].

REFERENCES

1. ISO/IEC 80079-20-2:2016 Explosive atmospheres - Part 20-2: Material characteristics – Combustible dusts test methods.
2. Pilz GmbH & Co. KG, 2017. *The Safety Compendium.* 5th ed. Pilz GmbH & Co. KG, Ostfildern, Germany.
3. IEC 60079-0:2017 "Explosive atmospheres – Part 0: Equipment – General requirements".
4. Amyotte P.R., Khan F.I., 2019. *Methods in Chemical Process Safety – Volume Three – Dust Explosion.* Academic Press, Elsevier, Cambridge, MA.
5. Europe Union, 2022. Summaries of EU legislation – Articolo 288 del trattato sul funzionamento dell'Unione europea – Direttive. https://eur-lex.europa.eu/IT/legal-content/summary/european-union-directives.html, last accessed 01/15/2026.
6. Europe Union, 2023. Regolamenti, direttive e altri atti, January 2023. https://european-union.europa.eu/institutions-law-budget/law/types-legislation_it, last accessed 01/15/2026.
7. CEN Cenelec, 2025. European Standards, European Committee for Standardization (CEN) and the European Committee for Electrotechnical Standardization (CENELEC) December 2025. https://www.cencenelec.eu/european-standardization/cen-and-cenelec/, last accessed 01/15/2026.
8. Amico A., Amico G., Bellomia G., 2002. Meccanica dell'incendio e valutazione del rischio. Flaccovio Ed.
9. EN 14491:2012 "Dust explosion venting protective systems".
10. EN 14797:2007 "Explosion venting devices".
11. EN 14460:2018 "Explosion resistant equipment".
12. Boring L.R., Roth E., Straeter O., Laumann K., Blackman S.H., Oxstrand J., Persensky J.J., 2009. Is human reliability relevant to human factors?. Proceedings of the Human Factors and Ergonomics Society 53rd Annual Meeting.
13. EN 15089:2009 "Explosion isolation systems".

14. Cloney C., 2025. NFPA 660 standard for combustible dust and particulate solids proposed. dustsafetyscience.com, March 2025. https://dustsafetyscience.com/nfpa-660-standard-for-combustible-dust/, last accessed 01/15/2026.
15. NFPA 660:2025 "Standard for combustible dusts and particulate solids".
16. NFPA 68:2023 "Standard on explosion protection by deflagration venting".
17. NFPA 69:2024 "Standard on explosion prevention systems".
18. NFPA 70:2023 "National electrical code".
19. NFPA 77:2024 "Recommended practice on static electricity".
20. NFPA 499:2024 "Recommended practice for the classification of combustible dusts and of hazardous (Classified) locations for electrical installations in chemical process areas".
21. United Nations Economic Commission for Europe, UNECE, 2023. Risk assessment for industrial accident prevention. United Nations, Geneva.
22. ISO 31000:2018 "Risk management – Guidelines".
23. Lark J., 2015. ISO 31000: Risk management – A practical guide for SMEs, ISO, ITC, UNIDO.
24. Yoe C., 2019. *Principles of Risk Analysis – Decision Making Under Uncertainty.* 2nd ed. Taylor & Francis Group LLC, Boca Raton, FL.
25. Crispin G., 2020. The essence of risk identification in project risk management: An overview. *International Journal of Science and Research.* DOI: 10.21275/SR20215023033.
26. IEC 31010:2019 "Risk management – Risk assessment techniques".
27. Poyhonen M., 2000. POLSSS: Surveying stakeholders about acceptability of risks and system changes. *Safety Science*, 35, 123--137.
28. Aven T., Zio E., 2018. *Knowledge in Risk Assessment and Management.* John Wiley & Sons Ltd, Oxford, UK.
29. Cameron I.T., Raman R., 2005. *Process Systems Risk Management.* Elsevier Inc., San Diego, CA.
30. Giusto C., 2019. *The Comprehensive Guide to Dust Hazard Analysis and OSHA Compliance.* Hallam-ICS, New York.
31. Cloney C., 2025. What is a Dust Hazard Analysis (DHA) and 10 important steps. dustsafetyscience.com, February 2025. https://dustsafetyscience.com/dust-hazard-analysis/, last accessed 01/15/2026.
32. Cloney C., 2024. 2023 combustible dust incident report - Version #1. DustEx Research Ltd. http://dustsafetyscience.com/2023-Report, last accessed 01/15/2026.
33. Eckhoff R.K., 2006. Differences and similarities of gas and dust explosions: A critical evaluation of the European "ATEX" directives in relation to dusts". *Journal of Loss Prevention in the Process Industry*, 19, 553–560.
34. Ogle A.R., 2017. *Dust Explosion Dynamics.* Elsevier Inc, Oxford, UK.
35. Piccinini N., 2019. *Analisi di Esplosioni e Incendi.* EPC S.r.l.
36. Mayers T.J., Ibarreta A.F., 2013. Tutorial on combustible dust. *Process Safety Progress*, 32(3), 298–306.
37. Turns S.R., 2012. *An Introduction to Combustion: Concepts and Applications.* 3rd ed. McGraw-Hill, New Delhi, India.
38. Eckhoff R.K., 2003. *Dust Explosions in the Process Industries.* 3rd ed. Gulf Professional Publishing (Elsevier Science), New York.
39. Cuccurullo G., 2025. *Elementi di Termodinamica e trasmissione del calore con prime applicazioni agli impianti termotecnici.* 3rd ed. Maggioli Editore.
40. French S., Bedford T., Pollard S.T.J., Soane E., 2011. Human reliability analysis: A critique and review for managers. *Safety Science*, 49(6), 753–763.
41. Grassani E., 2003. *L'errore come causa di infortunio.* Ed. UTET.

42. Gambetti F., Casalli A., Chisari V., 2012. The human factor in process safety management. *Chemical Engineering Transaction*, 26, 279–284.
43. Nolan D.P., 2014. *Handbook of Fire and Explosion Protection Engineering Principles.* 3rd ed. Elsevier, Amsterdam, The Netherlands.
44. Lehto M., Landry S.J., 2013. *Introduction to Human Factors and Ergonomics for Engineers.* 2nd ed. CRC Press Taylor & Francis Group, Boca Raton, FL.
45. Gunda R.Y., Gupta S., Singh K.L., 2023. Assessing human performance and human reliability: A review. *International Journal of System Assurance Engineering and Management*, 14(3), 817–828.
46. Dhillon S.B., 2014. *Human Reliability, Error, and Human Factors in Power Generation.* Springer International Publishing Cham, Switzerland.
47. Yalcin E., Altun Ciftcioglu G., Guzel B., 2024. Human reliability analysis methods. *Pamukkale University Journal of Engineering Sciences*, 30(2), 282–292.
48. Granger Morgan M., Fischoff B., Bostrom A., Atman J.C., 2002. *Risk Communication – A Mental Models Approach.* Cambridge University Press, Cambridge, UK.
49. OSHA Publications Office, 2009. Hazard communication guidance for combustible dusts, OSHA 3371-08.2009.
50. Hansson T.D., 2005. Risk analysis and a study of risk awareness and risk communication at LEAF Gävle. Report 5174. Brandteknik, Lunds tekniska högskola, Lunds universitet, Lund.
51. Mañez M., Carmona M., Haro D., Hanger S., 2016. Risk perception. In: J. Aerts & J. Mysiak, *Novel Multi-Sector Partnerships in Disaster Risk Management.* Results of the ENHANCE Project, 51–67.

6 Common Equipment and Processes in the Pharma and Food Industries

6.1 POWDERS AND EQUIPMENT OF THE PHARMACEUTICAL INDUSTRY

The pharmaceutical industry is fundamentally a powder-based manufacturing sector (even when the final product is in the liquid state; this is because many pharmaceutical liquids are solutions derived by mixing a powder with a given solvent). Particularly, the most relevant classes of dusts frequently handled and processed in pharmaceutical plants are:

- Active Pharmaceutical Ingredients
  An Active Pharmaceutical Ingredient (API) is the biologically active component of a medicinal product that is responsible for its therapeutic effect. Unlike excipients, which are pharmacologically inactive and serve supportive roles (e.g., dilution, stabilization, delivery), the API directly interacts with biological targets to diagnose, cure, mitigate, treat, or prevent disease. From a regulatory standpoint (EMA, FDA, ICH), an API is defined as: "Any substance or mixture of substances intended to be used in the manufacture of a pharmaceutical dosage form and that, when used, becomes an active ingredient of that dosage form." APIs may be synthetic, semi-synthetic, biotechnologically derived, or extracted from natural sources, and they can be administered in solid, liquid, or gaseous dosage forms. APIs span a wide range of molecular weights, functional groups (most of them classified as thermally unstable), solid-state forms (crystalline, amorphous, polymorphic, etc.), thermal and mechanical sensitivities. APIs are often produced via multi-step chemical synthesis (batch or flow), fermentation or cell culture, and extraction/purification from natural matrices. They are typically isolated as fine (d_{50} often minor of 50 μm), low-density and electrostatically active powders; therefore, numerous APIs qualify as combustible dusts, even when they are not flammable in their bulk form. Typical properties observed for organic APIs include a minimum ignition energy often lower than 30 mJ, a deflagration index commonly comprised of between 50 and 200 bar·m/s and a minimum explosible concentration of some tens of g/m^3. These properties must be considered during

DOI: 10.1201/9781003645320-6

scale-up and equipment design because they have implications in terms of explosion risk, especially during drying, milling, and handling operations. Common examples of Active Pharmaceutical Ingredients are paracetamol (acetaminophen), ibuprofen, aspirin (acetylsalicylic acid), diclofenac, naproxen, prednisolone, metoprolol, amlodipine, atorvastatin, diazepam, amoxicillin, ciprofloxacin, azithromycin, and many others.

- Excipients

  Excipients are pharmacologically inactive substances formulated together with the Active Pharmaceutical Ingredients. They enable manufacturing, stability, bioavailability, and patient acceptability of the dosage form. They can be classified into:

  (1) Fillers and diluents. They are used to add bulk and improve content uniformity. Some examples are: Lactose monohydrate, Microcrystalline Cellulose (MCC), Dicalcium phosphate, Mannitol, Starch, etc. Most of them are combustible dusts, exhibiting a typical K_{st} value ranging from 50 to 150 bar·m/s. MIE is often quite low (10–50 mJ). Such dusts are widely involved in silo, milling, blending, and conveying explosions.

  (2) Binders. They promote particle adhesion during granulation processes. Common examples are: Polyvinylpyrrolidone (PVP), Hydroxypropyl methylcellulose (HPMC), Gelatin, Pregelatinized starch, Sucrose syrup (wet granulation), etc. Dry polymeric binders (PVP, HPMC) are low-MIE organic powders, while gelatin and starch derivatives are thermally sensitive and combustible.

  (3) Disintegrants. They facilitate tablet breakup after administration. Some examples are: croscarmellose sodium, crospovidone, sodium starch glycolate, low-substituted hydroxypropyl cellulose (L-HPC), etc. As they generally are highly porous, if finely milled, they can exhibit a low MEC, making them, therefore, involved in fluid bed granulation and blending hazards.

  (4) Lubricants. They are used to reduce friction during tablet compression and ejection. Common examples are magnesium stearate, calcium stearate, stearic acid, and sodium stearyl fumarate. In general, fatty acid salts are combustible and electrostatically active. Fine magnesium stearate dust can exhibit quite low MIE values (up to 10 mJ) and has very low bulk density, increasing dispersion risk.

  (5) Glindants. They are used to improve powder flow. Common examples are colloidal silicon dioxide and talc. Such dusts are non-combustible; however, when mixed with organic excipients, they may not suppress explosibility. Therefore, their hybrid dust mixtures still require testing.

  (6) Coating Agents. They are generally used for taste masking, protection, or modified release. Common examples are hydroxypropyl methylcellulose (HPMC), ethylcellulose, Polyvinyl alcohol (PVA), and Shellac. Dry coating powders are all combustible; in particular, ethylcellulose dust exhibits a low MIE and a moderate K_{st}.

(7) Sweeteners. They improve palatability, especially in oral dosage forms. Common examples are sucrose, sorbitol, aspartame, and xylitol. Sugars are classical combustible dusts, and their explosions are well-documented historically.
(8) Colorants. They are used for identification and aesthetic purposes. Some examples include titanium dioxide (TiO_2) (non-combustible), iron oxides, and FD&C dyes (adsorbed on carriers). In general, pure inorganic pigments are non-combustible, but adsorbed dyes on organic carriers may be explosible as mixtures.
(9) Preservatives and Antioxidants. They basically improve the shelf life of the formulation. Common examples include parabens, butylated hydroxytoluene (BHT), and ascorbic acid. Many antioxidants are organic, low-MIE powders. For example, ascorbic acid dust is combustible and prone to thermal degradation.
(10) Modified-Release Polymers. They are used to control drug release kinetics. Common examples are ethylcellulose, methacrylates, and cellulose acetate. Polymeric excipients are thermoplastic and combustible; therefore, their molten behavior can affect venting and flameless vent suitability.

- Intermediates
  A pharmaceutical intermediate is a chemical substance that is formed during one or more steps of the synthesis of an API, and that is not intended for administration to patients. Intermediates serve as precursors or building blocks that are chemically transformed in subsequent processing steps to produce the final API. From a regulatory and process standpoint, a pharmaceutical intermediate can be defined as "a material produced during the manufacture of an API that undergoes further chemical transformation and does not yet meet the chemical, purity, or regulatory requirements of the final active substance." Intermediates may be isolated and stored or consumed immediately in the next synthesis step, depending on process design. From a process safety standpoint, pharmaceutical intermediates must be treated with the same rigor as APIs because: (a) many intermediates are organic, finely divided solids; (b) they may have lower MIE or higher K_{st} than the final API; (c) crystallization and drying steps often produce highly explosible powders; (d) some intermediates are thermally unstable or reactive. Typical hazards include dust explosions during drying, milling, or conveying operations, thermal decomposition, chemical reactivity, or incompatibility. As emphasized in FM Global and NFPA guidance, intermediates must be tested individually for dust explosibility parameters (MIE, MEC, K_{st}, P_{max}), as their behavior cannot be inferred from the API alone.
- Finished dosage forms
  A final dosage form is the finished pharmaceutical product in which the API is combined with suitable excipients and processed into a specific physical form designed for safe, accurate, and effective administration to the patient. In regulatory terms (EMA, FDA, ICH), a final dosage form

can be defined as "the physical form of a medicinal product that contains the active substance(s) and excipients, manufactured and packaged in its final configuration, and intended for administration to humans or animals." Once produced, the final dosage form is no longer subject to further chemical transformation and is released for distribution after quality control testing. Final dosage forms can be present in both solid and liquid phases. Concerning solid forms, the most common categories involve tablets (immediate-release, modified-release, effervescent), capsules (hard gelatin, soft gelatin), powders, and granules for oral use. From a process and safety standpoint, final dosage form production involves (a) blending, granulation, drying, compression, and coating operations; (b) large-scale handling of combustible excipients and APIs; (c) the use of equipment such as mixers, fluid bed dryers, tablet presses, coaters, and dust collectors.

Even though the product is "final," dust explosion hazards remain significant, especially for solid oral dosage forms, due to the possible presence of low MIE excipients (e.g., lactose, MCC), a fine particle size distribution, and the use of confined processing equipment. Therefore, final dosage form manufacturing requires the same level of dust hazard analysis and explosion protection as API and intermediate production.

All these powders can be involved in dust explosion events as a function of their main characteristics such as chemistry, granulometry, and its distribution, humidity content, storage or handling or process temperature, degree of turbulence when moved or processed, confinement, dispersion, etc. In the following sections, the most commonly used equipment (and, consequently, processes) of the pharmaceutical industry will be briefly described paying particular attention to their dust explosion risks.

6.1.1 Raw Material Storage, Charging, and Discharging Systems

Pharmaceutical manufacturing generally begins with the storage and handling of raw materials, which may include intermediates, APIs, and excipients.

These materials are generally stored in:

- **Drums**. Generally, they are 50–250 L rigid, closed containers, typically made of stainless steel, high-density polyethylene (HDPE), or fiber, designed to comply with Good Manufacturing Practice (GMP) requirements and to protect pharmaceutical materials from contamination, moisture, light, and mechanical damage.
- **Bags**. In most cases, they are 10–200 L flexible containers, typically made of polymeric films or multilayer materials, designed to protect pharmaceutical materials from contamination and to allow safe handling, transfer, or temporary storage, often inside a secondary rigid container.

- **Intermediate bulk containers (IBCs)**. They are rigid or semi-rigid containers, whose size can vary between 300 and 1000 L, typically with a square or rectangular footprint, designed to handle intermediate quantities of solids or liquids, while ensuring product protection, containment, and traceability.
- **Bins**. They can be considered as 100–1000 L reusable, rigid containers, most often stainless steel, designed to contain solid pharmaceutical materials and to interface directly with process equipment, ensuring cleanability, containment, and traceability.
- **Silos**. They are stationary, high-capacity containers (5–500 m^3), typically vertical, designed to store bulk raw materials or intermediates under controlled, GMP-compliant conditions, with systems for environmental control, segregation, and metered discharge to downstream processes depending on scale.

Charging and discharging operations, into and from the previously detailed equipment, are performed using gravity flow, screw feeders (a rotating helical screw (auger) housed inside a trough or tube. As the screw rotates, it draws material from the inlet and pushes it forward, allowing accurate feeding of solids by adjusting the screw speed), vacuum transfer systems (it moves solid materials by creating a pressure difference between the source and the "under vacuum" destination), or dense-phase pneumatic conveying (pneumatic transport method in which bulk solids are conveyed through a pipeline at low gas velocity and high solids concentration, using compressed air or inert gas, to move material gently and efficiently).

Dust explosion hazards arise primarily during filling, emptying, and transfer operations, when dust clouds can be generated inside containers and transfer lines. As already detailed, many pharmaceutical powders exhibit low MIE values, often in the range of 1–30 mJ, making them highly sensitive to electrostatic discharges. In addition, particle sizes are typically well below 100 µm, favoring suspension and flame propagation. This leads to K_{st} values commonly in the range from 50 to 200 bar·m/s (that is, St-1 or St-2 explosibility classes).

Moreover, confinement (an essential element of the dust explosion pentagon) is inherent to silos, bins, and IBCs; while dispersion (another essential element for dust explosion to occur) is practically unavoidable during pneumatic either filling or discharge. Because of these two elements, dust concentrations can readily exceed the MEC (typically between 30 and 100 g/m^3). As a result, these systems often meet all criteria for an equipment explosion hazard.

Explosion venting is the most common protection method for silos and bins. *Vent panels* (thin, intentionally weak membranes installed on equipment which rupture at a predetermined pressure, allowing hot gases, flame, and pressure to be safely released to a designated area) must be sized based on K_{st} and P_{max} data for the selected powder, industrial vessel volume, and desired reduced explosion pressure (P_{red}) following dedicated standards. Whenever venting to the outside is not possible, flameless venting devices (which combine an explosion vent panel with a flame-arresting and cooling element, allowing pressure relief without external flame propagation) may be used for no-metal content pharmaceutical dusts. Moreover, explosion isolation devices, such as *rotary air locks* (mechanical devices used to transfer bulk solids between two areas at different pressures while maintaining an air seal), *fast-acting valves* (valves designed to open or close in a very short time, typically milliseconds, in order to isolate, protect, or control a process during abnormal or safety-critical events), or *chemical isolation systems* (active explosion protection method that stops the propagation of an explosion, that is flame and pressure, through pipes or ducts by injecting a fast-acting chemical suppressant when an incipient explosion is detected) are required on both inlet and outlet connections to prevent flame propagation to upstream or downstream equipment.

In processes involving highly reactive APIs or hybrid mixtures (dust–solvent vapor systems), inerting is often applied. Nitrogen blanketing reduces oxygen concentration below the LOC, thereby eliminating the possibility of flame propagation even if ignition occurs. LOC values for pharmaceutical organic dusts are typically in the range of 10–15 %v/v O_2, but they must be experimentally determined for testing each material involved.

6.1.2 Size Reduction Equipment: Mills and Micronizers

Milling and micronization are central unit operations in pharmaceutical manufacturing, used to control particle size distribution and improve bioavailability. Common equipment includes:

- Hammer mills
  A hammer mill is a mechanical size-reduction device that reduces solid materials by impact, and to a lesser extent by shear and attrition. In the pharmaceutical industry, it is mainly used for dry milling of crystalline or granular materials to obtain a controlled particle size prior to blending, granulation, or tableting. A typical pharmaceutical hammer mill consists of the following elements (see Figure 6.1): a feed system composed of a feed hopper (introducing raw material into the mill by gravity or screw feeding), a grinding chamber (constituted by a rotor, the hammers (blades made of hardened stainless steel, typically AISI 316L for pharmaceutical applications), a stator, and a perforated screen which controls the maximum particle size of the product), a drive system constituted by an electric motor having a typical rotational speed of 2000–6000 rpm (depending on mill size), and a discharge system (often assisted by airflow) which collects the product into bins, bags, or other connected downstream equipment.

  The working principle is very simple. Solid particles enter the grinding chamber through the hopper. The rotating hammers accelerate the particles and repeatedly strike them at high velocity. The most important size reduction mechanisms are impact (hammer–particle collision), shear (particle–particle and particle–screen interactions), and attrition (friction against chamber walls and screen). Only particles smaller than the screen holes can exit the chamber; oversized particles remain inside for further

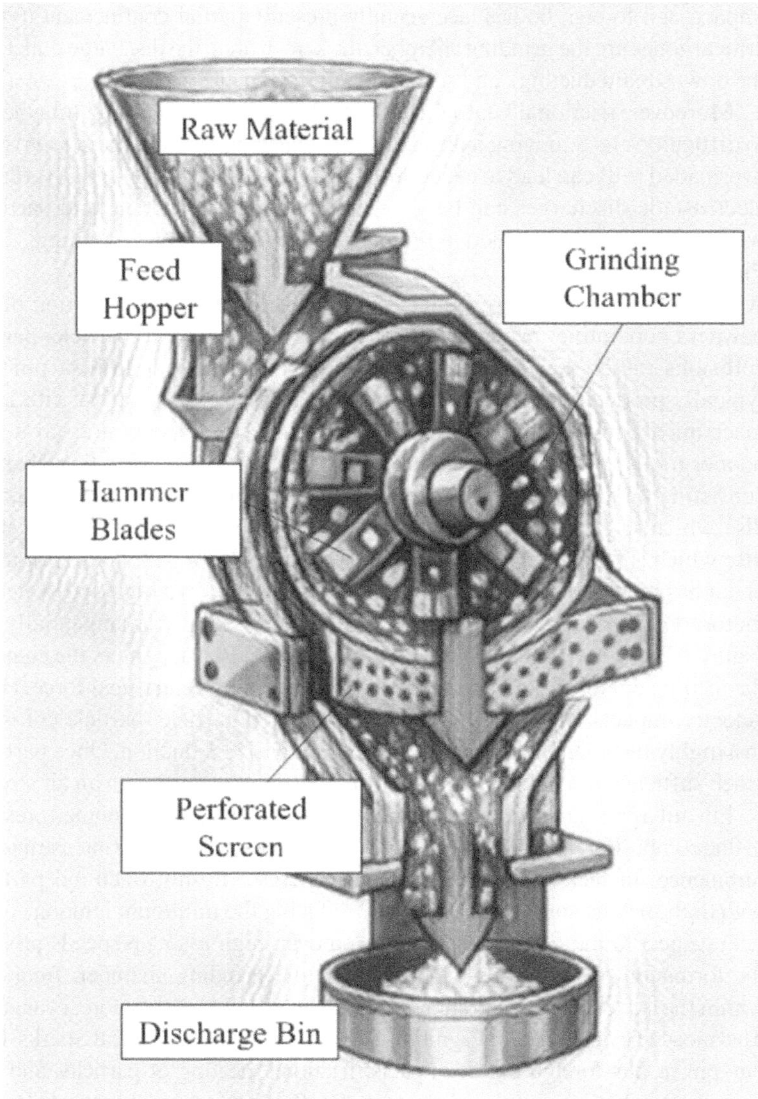

FIGURE 6.1 Schematic illustration of a typical pharma hammer mill.

milling. Finally, milled powder exits through the screen, and it is conveyed out by gravity (or air).

A hammer mill is intrinsically prone to dust explosion risk because it simultaneously provides combustible dust generation (with a fine particle size distribution which intrinsically implies lower MIE and higher K_{st} values with respect to the original material), ignition sources such as mechanical heat or sparks (generated by hammer–hammer contact (wear or failure), hammer–screen impact, and foreign bodies (accidentally present), partial confinement (typical critical zones are the grinding chamber, the screen area, the discharge outlet, and the downstream ducting), and suspension of dust in air.

Moreover, frictional heating due to prolonged residence time (maybe due to difficulties in achieving a dust capable of passing through the screen) or an overloaded mill can lead to exceeding the MIT of the powder. Finally, further electrostatic discharges can be generated by particle–particle and particle–wall charging, non-conductive liners or screens, and poor grounding.

- Pin mills

  A pin mill is a high-energy impact mill used for fine size reduction of dry powders, operating primarily through particle–pin and particle–particle collisions rather than crushing. Compared to a hammer mill, a pin mill typically produces a finer and narrower particle size distribution with lower mechanical stress on the equipment. Its main components are: (a) a feed hopper (or screw feeder) for controlled solids input; (b) a grinding chamber consisting of two circular discs fitted with multiple rows of pins; (c) a rotor disc which rotates at high speed and has pins mounted radially; (d) a stator disc which is fixed (single-rotor mills) or counter-rotating (double-rotor mills) and it has pins arranged to intermesh with rotor pins; (e) a high-speed electric motor; (f) a discharge system where product exits radially or tangentially (see Figure 6.2). The working principle is quite simple. Powder enters the center of the milling chamber, and it is accelerated outward by centrifugal force. High-velocity impacts with rotating pins, and repeated particle–particle collisions in a highly turbulent field provide the necessary size reduction. Once particles reach sufficient fineness and momentum, they exit the mill with an air stream.

  Pin mills present a significant dust explosion risk due to the combined presence of fine combustible powders, high-energy mechanical motion, and intense internal turbulence. In fact, the milling process generates freshly fractured particles with high specific surface area, strongly reducing the minimum ignition energy with respect to the original material. Moreover, high rotor tip speeds promote the formation of dense dust clouds within the grinding chamber, frequently within the explosive concentration range (MEC should be, therefore, evaluated). The most relevant potential ignition sources include mechanical sparks from pin–pin or pin–foreign object impacts, frictional heating of particles and mill components, hot bearings, and electrostatic discharges caused by particle–wall interactions, particularly under low-humidity conditions. Although pin mills are typically less confined than hammer mills, the grinding chamber and connected ductwork still allow pressure buildup and flame acceleration. An ignition within the mill can propagate to downstream equipment, such as cyclones or filters, increasing the risk of secondary explosions if adequate prevention and protection measures are not implemented.

- Ball mills

  A ball mill is a tumbling mill used for the size reduction and homogenization of solid materials, operating primarily through impact and attrition between grinding media and particles. In pharmaceutical applications, ball

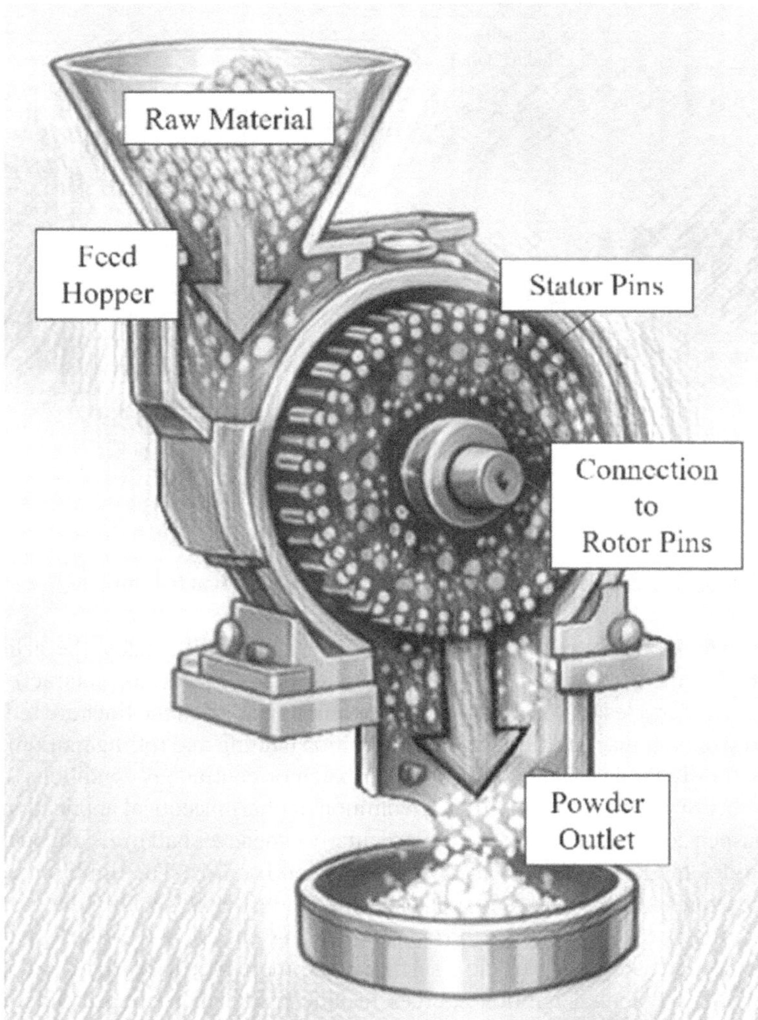

FIGURE 6.2 Schematic illustration of a typical pharma pin mill.

mills are mainly used for coarse to fine grinding, blending, and sometimes mechanochemical processing (often used to either synthesize APIs and final products or promote different crystallization paths). Its main components are (see Figure 6.3): (a) a milling chamber (which is essentially a cylindrical shell, usually horizontal, made of steel or stainless steel, internally lined with wear-resistant material as ceramic or polymer); (b) a grinding media, that is balls made of stainless steel, alumina, or zirconia with a typical diameter range between 1 and 50 mm; (c) an electric motor with gearbox which rotates the mill at a controlled speed; (d) a discontinuous feed and discharge system (batch emptying).

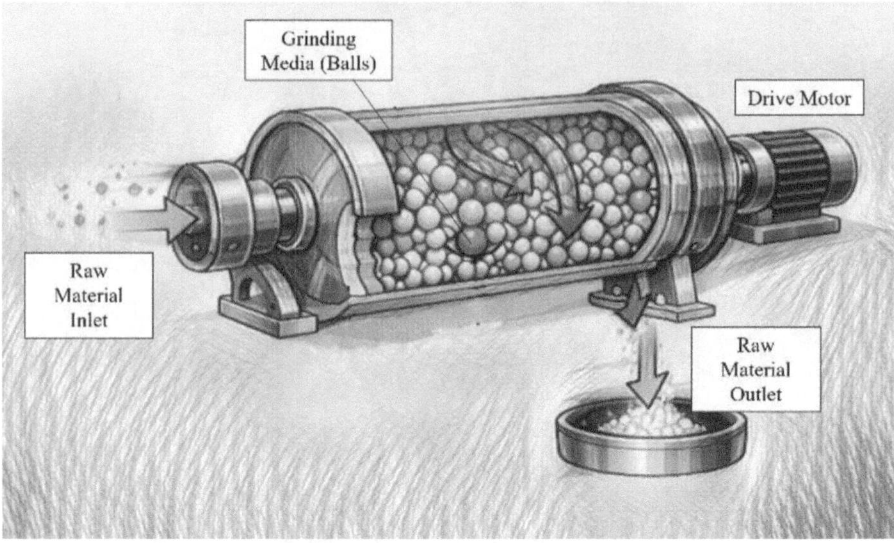

FIGURE 6.3 Schematic illustration of a typical pharmaceutical ball mill.

The working principle is very simple. As the mill rotates, the grinding media are lifted along the shell wall and then cascade (or cataract) back down. Particles trapped among balls or among balls and the liner are reduced in size by impact (falling balls), and attrition (sliding and rolling motion). The ball mill can be operated under either batch or continuous conditions, using either dry milling or wet milling (common in pharmaceutical applications for suspensions and nano-milling). Particularly, whenever ball mills are operated in dry mode, they can exhibit dust explosion hazards. The tumbling action generates fine particles and creates localized dust clouds within the partially confined rotating chamber. Prolonged milling unavoidably reduces particle size, increasing the particle's surface area, therefore, implying a reduction of MIE. Potential ignition sources include frictional heating at liners and grinding media contacts, hot bearings, and electrostatic charge accumulation (particularly with non-conductive liners or low humidity). Although ball mills typically operate at lower tip speeds than pin mills, long residence times and enclosed volumes can allow pressure buildup. An internal ignition may propagate to connected equipment, posing risks of secondary explosions without adequate inerting, grounding, and explosion protection measures.

- Jet mills

  A jet mill (also called a fluid energy mill) is a fine grinding device that reduces particle size through high velocity particle–particle collisions induced by compressed gas jets. Unlike mechanical mills, jet mills contain no moving parts in the grinding zone, making them particularly suitable for ultrafine and contamination-sensitive pharmaceutical materials. Its main components are: (a) a deed injector which introduces powder into the milling chamber; (b) a grinding chamber, typically circular or toroidal;

(c) gas nozzles which inject compressed air, or inert gas, at high velocity (often supersonic); (d) an internal classifier (sometimes optional), rotating or static, which controls top particle size; (e) a discharge outlet which generally conveys fine particles to a cyclone (see Figure 6.4).

The working principle is quite simple. Compressed gas passing through the nozzles generates high velocity jets that accelerate particles to very high speeds. Size reduction occurs primarily through interparticle impact in the turbulent grinding zone. Finer particles are carried out of the mill, while coarser particles remain until further reduced. In classifier jet mills, only particles below a target cut size can exit.

Jet mills present specific dust explosion hazards due to the intentional formation of highly dispersed, fine powder clouds within a confined grinding chamber. The high turbulence generated by compressed gas jets promotes dust concentrations above MEC, while micronized particles exhibit reduced MIE and higher K_{st} with respect to the original material. Although jet mills lack mechanical moving parts in the grinding zone, potential ignition sources include electrostatic discharges from particle–wall interactions, and frictional heating at nozzles. When operated with air rather than inert gas, an internal ignition may propagate to downstream cyclones or filters, creating a risk of secondary explosions if appropriate inerting, grounding, and explosion protection measures are not installed.

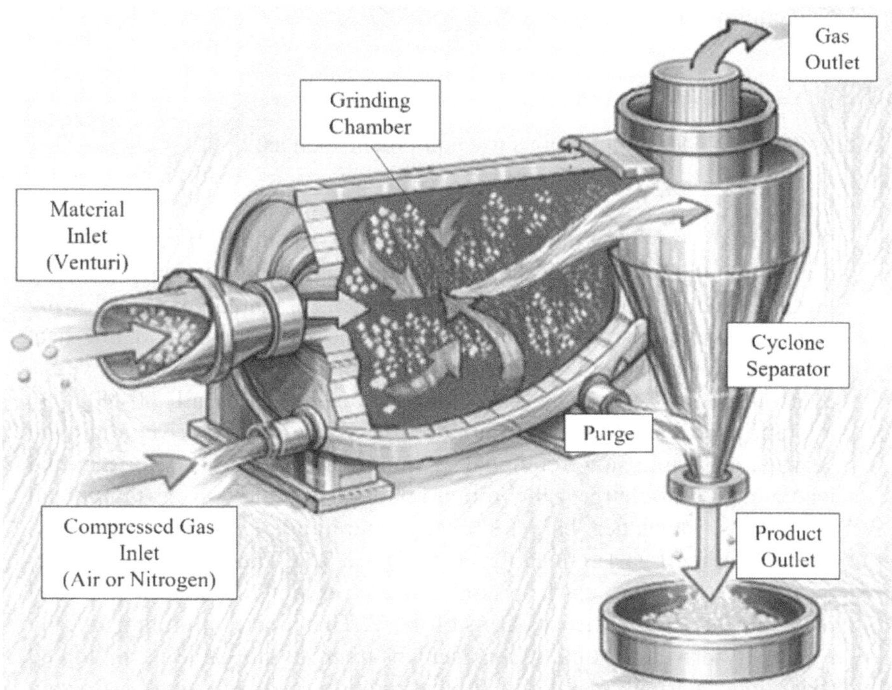

FIGURE 6.4 Schematic illustration of a typical pharmaceutical jet mill.

Summarizing, all the units devoted to decrease the particle size of a given material significantly increase dust explosion risk during processing because they intentionally apply significant mechanical energy to particles, also promoting the formation of dispersed dust clouds. Particle size reduction increases dust particles' specific surface area and often reduces moisture content (which can either vaporize or bind into the dust), both of which unavoidably lower the MIE and increase the K_{st} with respect to the original product. Jet-milled APIs, for example, may exhibit MIE values below 5 mJ and K_{st} values exceeding 150 bar·m/s. Mills inherently generate dense dust clouds inside confined housings, frequently well above the MEC. Credible ignition sources include mechanical friction, metal-to-metal contact, foreign objects, overheated bearings, and electrostatic discharges. As a result, mills are among the highest risk pieces of equipment in pharmaceutical plants.

Explosion venting is commonly applied to milling equipment, with vents directed to a safe outdoor location. For indoor installations where vent ducts are impractical, flameless venting devices may be used, provided the dust is compatible. For high-value or "mission-critical" mills, explosion suppression systems are often selected. These systems detect the early pressure rise of a deflagration and inject suppressant to limit P_{red} below the vessel design pressure.

Finally, inert gas milling is widely used for APIs with extremely low MIE or when solvent vapors are present. By maintaining oxygen concentration below the LOC, the formation of an explosible atmosphere is therefore prevented. Moreover, explosion isolation should be mandatory at both the mill inlets and outlets to protect connected equipment.

6.1.3 Mixers and Blenders

Pharmaceutical blending operations use equipment such as:

- V-blenders
  A V-blender (or V-type tumble blender) is a low-shear, batch-mixing device used in the pharmaceutical industry for the homogeneous blending of dry powders and granular materials. It is especially suited for free-flowing pharmaceutical dusts where minimal particle size change and low heat generation are required. Its main components are: (a) a V-shaped vessel composed of two cylindrical shells joined at an angle (typically 60–90°); (b) a rotating shaft supported by trunnions (or bearings); (c) a drive motor and gearbox providing slow rotational motion; (d) charging ports for material loading; (e) a discharge valve (often butterfly or slide valve); (f) optional devices as intensifier bar (for deagglomeration), sampling ports, and vacuum (or inert gas) connections (see Figure 6.5). The V-blender operates by tumbling motion: as the vessel rotates, powder repeatedly splits and recombines between the two arms of the V. This creates mixing primarily by gravity-driven particle rearrangement (there is no grinding or forced shear under normal operation). Typical rotation speed is around 5–25 rpm (low rotational speed limits heat generation and mechanical stress) with a

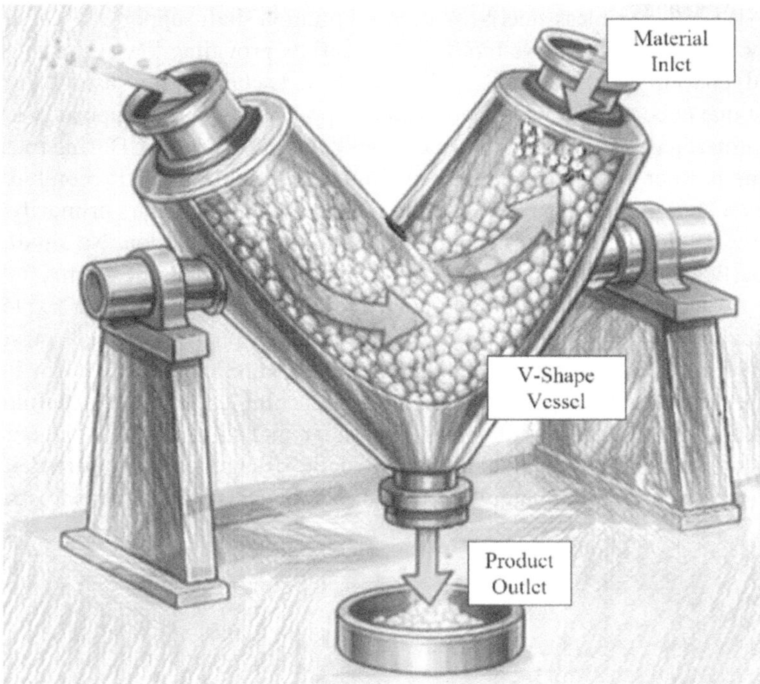

FIGURE 6.5 Schematic illustration of a pharmaceutical V-blender.

fill level of 40–70% of the total volume. The mixing time can vary from a few minutes to tens of minutes, depending on material properties.

V-blenders present a potential dust explosion hazard when handling fine, combustible pharmaceutical powders. During charging, tumbling, and discharge, dust clouds can form inside the vessel or in the surrounding area, particularly at low fill levels and high rotation speeds. Although the blending process itself is low shear and generates minimal heat, ignition sources may include electrostatic discharges from particle–wall interactions, inadequately grounded equipment, and external sources during loading or unloading. The enclosed geometry of the blender can allow pressure buildup if ignition occurs. Additionally, dust accumulation in seals, bearings, and nearby surfaces increases the risk of secondary explosions. Proper grounding, inerting when required, controlled filling, and effective housekeeping are essential to mitigate these risks.

- Double-cone blenders

  A double-cone blender is a "tumble type" batch mixer used in the pharmaceutical industry for gentle and homogeneous blending of dry powders and granules. It consists of two truncated cones joined at their bases, forming a symmetrical vessel that promotes efficient powder movement with minimal shear. Its main components are: (a) a double-cone vessel (typically of

AISI 316L stainless steel); (b) a central rotation shaft supported by bearings or trunnions; (c) a drive motor and a gearbox providing low-speed rotation; (d) charging ports for material loading; (e) a discharge valve, usually located at one cone apex; (f) some optional components as sampling ports, vacuum or inert gas connections, and intensifier bar (see Figure 6.6). During rotation, the powder bed undergoes repeated tumbling, splitting, and recombination as material flows from one cone to the other. Mixing occurs primarily by a gravity-driven mechanism rather than mechanical agitation. No significant particle breakage or size reduction occurs under normal operation.

As for all mixers, double-cone blenders may present dust explosion hazards when handling fine, combustible pharmaceutical powders. During charging, tumbling, and discharge, dust clouds can form inside the vessel and at openings, potentially reaching explosive concentrations. Although the process involves low shear and minimal heat generation, ignition sources may include electrostatic discharges from particle–wall interactions, inadequately grounded components, and external sources

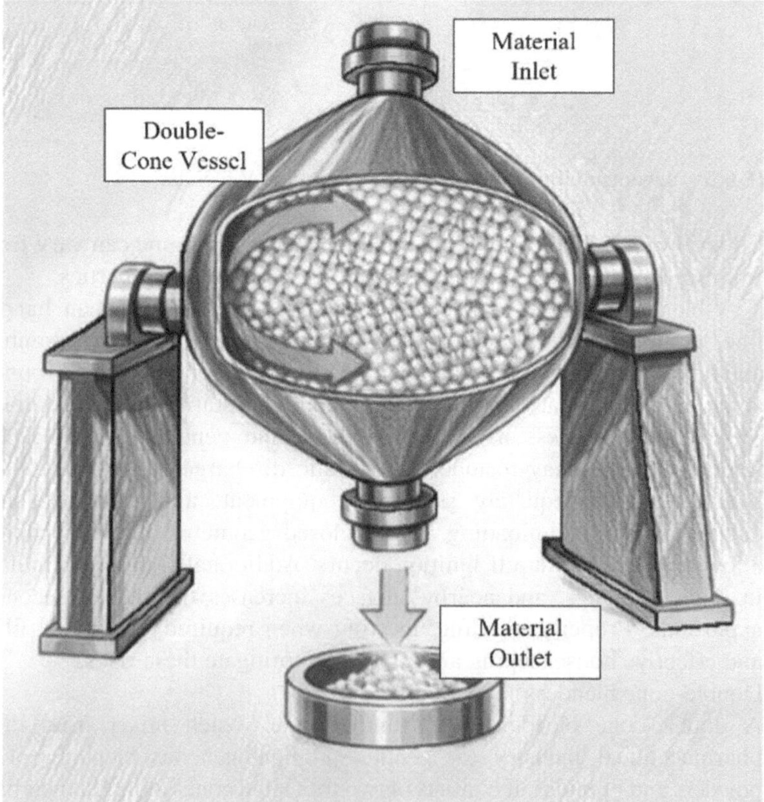

FIGURE 6.6 Schematic illustration of a double-cone blender.

during loading or unloading. Even in this piece of equipment, the enclosed geometry can allow pressure buildup if ignition occurs. Proper grounding, controlled filling, inerting when necessary, and effective housekeeping are always essential to mitigate dust explosion risks.

- Ribbon mixers
  A ribbon mixer is a horizontal, mechanically agitated batch mixer used in the pharmaceutical industry for the homogeneous mixing of dry powders, granules, and low-viscosity pastes. It provides moderate shear mixing, making it suitable for cohesive powders and formulations requiring faster blending than tumble blenders. Its main components are (see Figure 6.7): (a) a U-shaped horizontal trough (typically AISI 316L stainless steel); (b) a central rotating shaft fitted with an outer helical ribbon (which moves material in one axial direction) and an inner helical ribbon (which moves material in the opposite direction); (c) end bearings and seals; (d) a drive motor and a gearbox providing controlled rotational speed; (e) top charging ports for material loading; (f) a bottom discharge valve (usually full-length or center-mounted). Optional features include liquid spray nozzles, heating/cooling jackets, and vacuum or inert gas connections. During operation, the inner and outer ribbons rotate together, creating counter-current axial and radial powder movement. Material is continuously lifted, folded, and redistributed across the vessel, resulting in rapid and uniform mixing

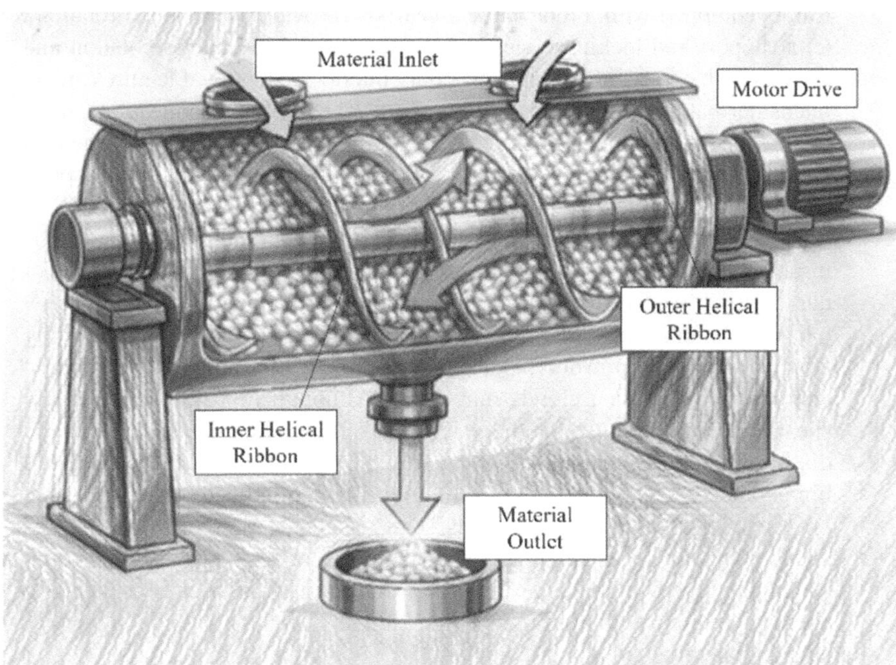

FIGURE 6.7 Schematic illustration of a pharmaceutical ribbon mixer.

mainly through convective and shear mechanisms. Rotation speed is typically around 20–100 rpm, fill level is of 40–70% of vessel volume with a mixing time of minutes (depending on formulation).

Likeall the other blenders, ribbon mixers exhibit a dust explosion hazard whenever processing fine, combustible powders due to the generation of suspended dust clouds during charging, mixing, and discharge. The rotating ribbons introduce mechanical energy, frictional heating, and potential metal–metal contact, which may act as ignition sources. In addition, electrostatic charge accumulation can occur if powders or internal surfaces are poorly grounded. The partially enclosed trough allows pressure buildup if ignition occurs, while dust accumulation in seals, bearings, and surrounding areas increases the risk of secondary explosions. Proper grounding, inerting, controlled operation, and effective housekeeping are, even in this piece of equipment, essential to mitigate these risks.

- Bin blenders

  A bin blender (also called a container blender) is a tumble-type batch mixing system in which powders are blended directly inside a removable process container (bin). The same bin can be used for charging, blending, transport, and discharge, minimizing powder handling and cross-contamination. Its main components are: (a) a process bin, typically of stainless steel (AISI 316L), of cylindrical (conical, or rectangular) geometry equipped with charging ports and a discharge valve; (b) a blender drive unit (rotating or tumbling frame) equipped with a motor and a gearbox (providing controlled rotation); (c) a support and locking system which secures the bin during rotation and ensures both correct alignment and safety interlocks. Optional features are an intensifier bar (for deagglomeration), vacuum or inert gas connections, dust-tight seals, and contained interfaces (see Figure 6.8). The bin blender operates by tumbling motion. As the container rotates, the powder bed repeatedly splits, cascades, and recombines, promoting mixing. Mixing occurs with very low shear, and no internal moving parts contact the powder unless an intensifier is installed. Typical rotational speeds are around 5–15 rpm, with a filling level of dust of 40-70% of the total bin volume and residence times of some minutes.

  Bin blenders present dust explosion hazards whenever handling fine, combustible powders, particularly during charging and discharge operations where dust clouds may form. Although mixing occurs at low shear and generates minimal heat, ignition sources can include electrostatic discharges, poorly grounded bins, and external equipment. The enclosed bin geometry may allow pressure buildup if ignition occurs, with possible propagation to connected systems. Dust accumulation on bin surfaces and nearby equipment increases the risk of secondary explosions. Proper grounding, containment, inerting when required, and strict housekeeping are always essential risk mitigation measures.

- High-shear mixers

  A high-shear mixer (often called a high-shear granulator) is an intensive batch mixing device used in the pharmaceutical industry for rapid blending,

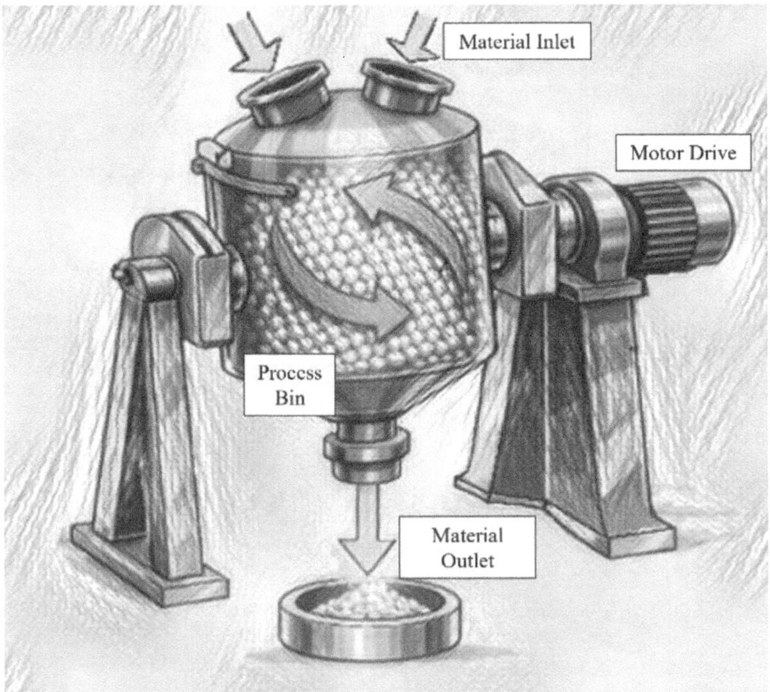

FIGURE 6.8 Schematic illustration of a pharmaceutical bin blender.

wet granulation, and agglomeration of powders. It operates by imparting high mechanical and shear energy to the powder bed, promoting fast and uniform mixing. Its main components are: (a) a mixing bowl (vessel), cylindrical or conical (typically of AISI 316L stainless steel), equipped with a product discharge port; (b) an impeller (main mixer) mounted at the bottom (or side) of the vessel which provides bulk mixing and eventual densification; (c) a chopper, that is a high speed secondary blade which breaks agglomerates and controls granule size; (d) drive motors and gearboxes (independent drives for impeller and chopper); (e) a liquid addition system with spray nozzle for the binder solution. Optional features include vacuum operation, jacketed vessels for temperature control, and automated cleaning (see Figure 6.9). Powders are charged into the vessel and mixed by the rotating impeller. During wet granulation, a binder solution is sprayed onto the moving powder bed. The combined action of the impeller (mixing) and the chopper (high local shear) promotes eventual rapid granule formation, densification, and size control. Typical impeller speed ranges between 50 and 500 rpm, with a chopper speed of up to several thousand rpm. This results in a high mechanical energy inside the vessel, together with a significant shear and heat generation.

High-shear mixers pose a significant dust explosion risk due to the simultaneous presence of fine combustible powders, intense mechanical

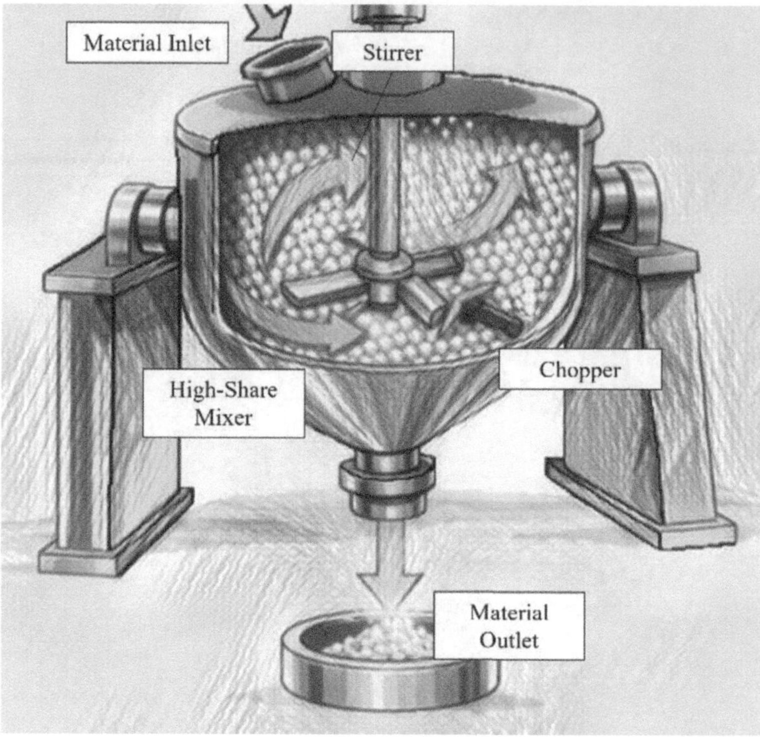

FIGURE 6.9 Schematic illustration of a pharmaceutical high-shear mixer.

energy, and partial confinement. High impeller and chopper speeds generate turbulence, frictional heating, and potential mechanical sparks, while rapid powder dispersion can produce dust concentrations within the explosive range. Electrostatic charging and hot surfaces further increase ignition likelihood. An internal ignition may result in a rapid pressure rise and flame propagation to connected equipment. Effective grounding, inerting, explosion protection, and strict control of operating conditions are, as always, essential to mitigate these hazards.

Summarizing, these units are typically batch-operated and handle large quantities of powder. Dust clouds are generated during charging, mixing, and discharging. Electrostatic charging is common due to particle–particle and particle–wall interactions. Many excipients and APIs exhibit low MIE, making electrostatic ignition a dominant risk. Dust concentrations inside mixers can exceed the MEC, particularly during tumbling or high-shear operation.

Explosion venting panels are often installed on mixer lids or sidewalls. In contained or cleanroom environments, flameless venting or explosion-resistant design may be preferred. Inerting with nitrogen is frequently used for high-shear mixers, particularly when processing fine, dry powders. Explosion isolation valves or chemical barriers are required on connected ducts or vacuum transfer lines.

6.1.4 Dryers and Granulators

In the pharmaceutical industry, several types of dryers are employed depending on product sensitivity and process requirements. Tray dryers and vacuum tray dryers are used for batch drying of heat-stable and heat-sensitive solids, respectively. Dynamic vacuum dryers provide gentle mixing under reduced pressure, improving uniformity. Fluid bed dryers enable rapid and homogeneous drying of granules through intense gas–solid contact. Spray dryers allow simultaneous drying and particle formation from liquid feeds, often producing amorphous solids. Freeze dryers remove water by sublimation and are essential for thermolabile biological products requiring maximum stability. In the following, static tray, dynamic, and fluid bed dryers will be briefly described, highlighting their safety concerns regarding dust explosion risk. Spray and freeze dyers will be described in the section related to food industry equipment (see Section 6.2.4).

A static tray dryer is a batch drying unit used to remove residual moisture or solvent (e.g., methanol, water, etc.) from powders, granules, or wet solids by circulating heated air or by applying vacuum over a material spread in shallow layers on trays. The material remains stationary during drying, and moisture/solvent removal occurs by either convective heat or mass transfer (vaporization). Its main components are: (a) a drying chamber, that is an insulated enclosure (typically of AISI 304 or 316L stainless steel) designed for either uniform airflow distribution or readily moisture removal; (b) a series of perforated (or solid) stainless steel trays arranged on racks with defined spacing; (c) an air handling system constituted by a blower (or fan or vacuum pump), a trays heater (electric or steam), and eventually, inlet air filters (often HEPA for pharmaceutical use); (d) an exhaust system formed by dampers and an exhaust fan; (e) a system for temperature, airflow, and time control (see Figure 6.10). Wet material is loaded onto trays and placed in the drying chamber. Heated air is circulated (or vacuum is applied) across the tray surfaces (which are generally heated), evaporating moisture/solvents from the solids. The exhaust air (or vacuum) removes the vapor, while fresh heated air (vacuum pump) maintains the driving force for drying. Drying proceeds slowly and uniformly due to low turbulence. Internal temperatures typically vary between 30 and 120°C, with a drying time of hours and a dust layer thickness ranging from a few millimeters to centimeters (depending on either the type of powder or the required moisture/solvent removal efficiency).

Static tray dryers can present dust explosion hazards whenever they process fine, combustible pharmaceutical powders, even wet by flammable solvents (such as methanol, which is often used in pharmaceutical applications), both under air and vacuum operation. During loading, drying, and unloading, dust clouds may form, especially when handling dry, low-density materials. In air-operated dryers, the presence of oxygen combined with heated surfaces and airborne dust can lead to ignition if temperatures exceed the powder minimum ignition temperature (both MIT and LIT) or if electrostatic discharges occur. Vacuum tray dryers reduce oxygen concentration and generally lower explosion risk; however, air ingress, leaks, or improper inerting can reintroduce hazardous conditions. Additional ignition sources include hot heater surfaces, electrical components, and static charge accumulation

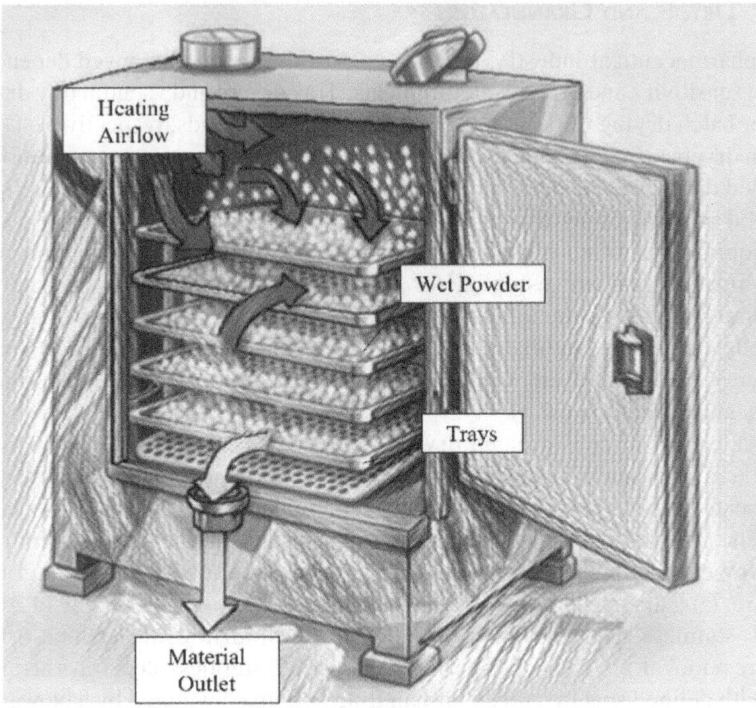

FIGURE 6.10 Schematic illustration of a static tray dryer.

on trays and racks. Effective mitigation requires grounding, temperature control, oxygen monitoring, inerting when appropriate, and careful operating and cleaning procedures. Moreover, further considerations should be introduced whenever there is the possibility of a hybrid mixture (e.g., dust and flammable solvent) formation.

A dynamic dryer is a batch drying unit designed to dry powders, granules, or pastes while they are continuously moved by an internal agitator. Unlike static dryers, the presence of mechanical agitation promotes uniform heat and mass transfer, reduces drying time, and prevents agglomeration or localized overheating. In pharmaceutical applications, dynamic dryers are often operated under vacuum to enable low-temperature drying of heat (or oxygen)-sensitive materials. Its main components are: (a) a drying vessel (cylindrical or conical), typically of AISI 316L stainless steel; (b) an internal agitator (paddle, ploughshare, anchor, or helical type); (c) a drive motor and a gearbox for agitator rotation; (d) a heating system (jacketed vessel and sometimes heated agitator shaft); (e) a vacuum system with condenser and receiver; (f) charging and discharge ports, often vacuum-tight (see Figure 6.11). Wet material is charged into the vessel, which is heated and evacuated if required. The agitator continuously mixes and turns the powder, renewing the surface exposed to heat and enhancing solvent evaporation. Vapor is removed through the vacuum line until the target residual moisture/solvent content is reached.

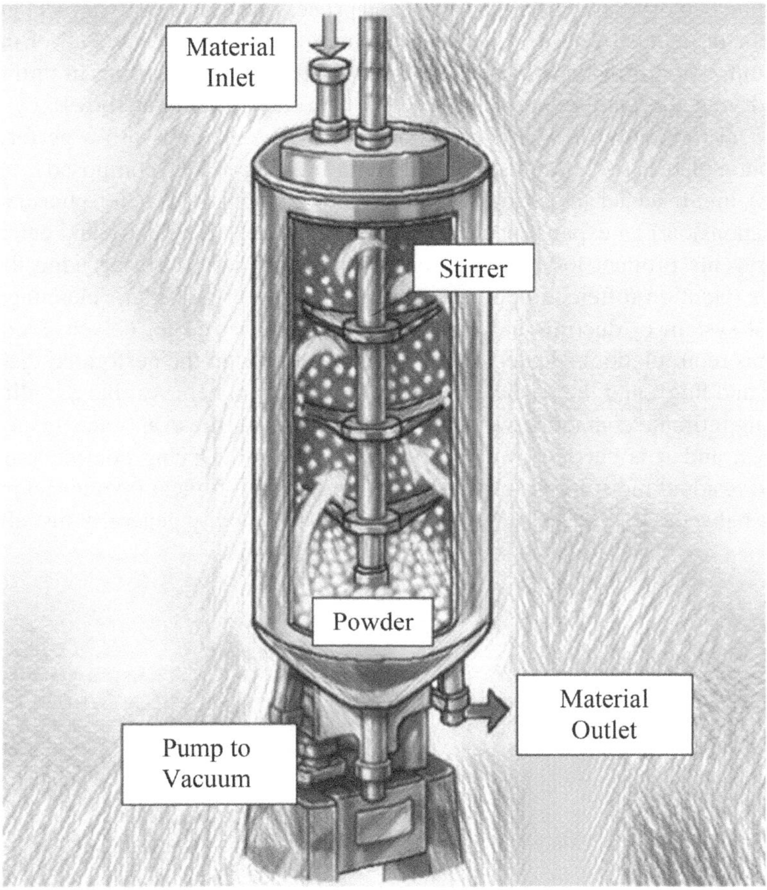

FIGURE 6.11 Schematic illustration of a dynamic dryer.

Dynamic vacuum dryers exhibit dust explosion hazards when processing fine, combustible powders due to the combination of mechanical agitation, partial confinement, and reduced moisture content during drying. Although vacuum operation lowers oxygen concentration and generally reduces explosion likelihood, air ingress through leaks or improper purging can reintroduce flammable atmospheres. The agitator can generate frictional heating, mechanical sparks, or electrostatic charging, particularly with cohesive or low-conductivity powders. Drying increases particle reactivity (most intermediate pharmaceutical compounds contain epoxide rings capable of opening and releasing a huge amount of heat) and lowers the minimum ignition energy. An ignition inside the vessel can result in a rapid pressure rise and propagation to condensers or receivers. Effective risk mitigation requires reliable inerting, oxygen monitoring, grounding, temperature control, explosion isolation, and strict maintenance of vacuum integrity.

A fluid bed dryer is a thermal drying and processing unit used to dry wet granules, powders, or pellets by suspending them in an upward flow of heated air or inert gas. The fluidization provides efficient heat and mass transfer, resulting in uniform and rapid drying. Its main components are: (a) a product container (bowl), cylindrical or conical (typically of AISI 316L stainless steel) equipped with a perforated air distributor plate at the bottom; (b) an Air Handling Unit (AHU) composed by a blower (or fan), an air heater (electric or steam), inlet air filters (HEPA for pharmaceutical applications); (c) an expansion chamber which reduces air velocity and entrainment and prevents product loss; (d) a filter housing with fabric or cartridge filters for powder retention (often equipped with filter shaking or pulsed air cleaning); (e) an exhaust system of ducting and fan (optionally customized for solvent recovery or inert gas recirculation). Heated air flows upward through the perforated distributor, lifting and fluidizing the particle bed. The wet material behaves like a boiling fluid, allowing intimate contact between gas and solids. Moisture evaporates from particle surfaces, and it is carried away with the exhaust air. Drying continues until the desired residual moisture is achieved (see Figure 6.12). Typical inlet air temperature ranges between 40 and 120°C, with superficial air velocity generally just above the

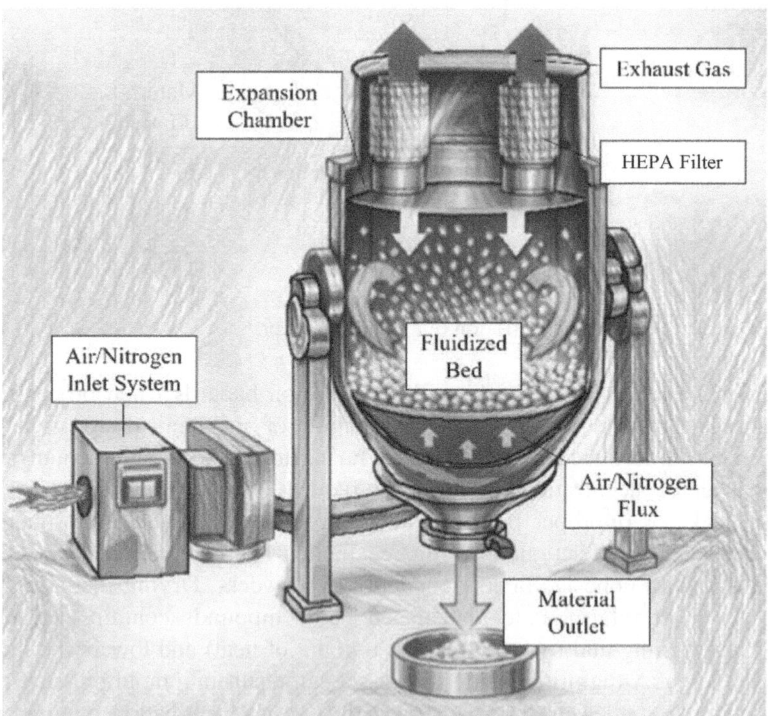

FIGURE 6.12 Schematic illustration of a fluid bed dryer.

minimum fluidization velocity. Drying time varies from minutes to tens of minutes, depending on the type of dust and the degree of humidity reduction required.

Fluid bed dryers present significant dust explosion hazards due to the intentional generation of highly dispersed, fine powder clouds under intense airflow and partial confinement. During operation, particles are fully suspended, and dust concentrations can readily fall within the explosive range, especially during start-up, shutdown, and discharge. Drying reduces moisture content, often increasing powder reactivity and lowering the minimum ignition energy. Potential ignition sources include electrostatic discharges from particle–particle and particle–wall interactions, hot inlet air, overheated heater elements, friction at distributor plates, and hot surfaces in fans or bearings. Moreover, if the powder is reactive, as most of epoxy dusts used as APIs intermediates, can react in correspondence of hot zones (e.g., the perforated plate or grids), providing a further source of ignition (smoldering).

Filter housings and expansion chambers represent critical confined volumes where dust accumulation and flame acceleration may occur. An ignition in the product bowl can propagate to exhaust ducts, cyclones, or filters, leading to secondary explosions with higher severity. Additionally, dust deposits on filters and surrounding equipment increase the risk of secondary events. Effective mitigation requires inert gas operation when appropriate, rigorous grounding, temperature control, explosion venting or suppression, explosion isolation, and strict operational and housekeeping procedures.

Another important pharmaceutical operation is granulation. Even if granulation generally implies increase in the dust particle size, granulation processes can involve several safety risks that must be carefully managed. During the early stages of the granulation, the huge amounts of fine powders can generate dust clouds with concentrations above MEC; therefore, the control of ignition sources and effective grounding are essential. Solvent-based wet granulation may introduce flammability risks, requiring proper inerting.

A fluid bed granulator is, generally, a multifunctional pharmaceutical processing unit used to perform wet granulation (with water), drying, and sometimes coating within the same equipment. It combines powder fluidization with controlled binder liquid spraying, producing granules with improved flowability, compressibility, and content uniformity. Its main components are: (a) a product container (granulation bowl), typically of AISI 316L stainless steel, equipped with a perforated air distributor plate; (b) an air handling unit (AHU) formed by a blower/fan, an air heater, and inlet HEPA filters; (c) a spray system (which can be top, bottom, or tangential) loading a binder solution by means of a pump connected to a tank; (d) an expansion chamber and a series of filters which prevent product entrainment; (e) a system of ducting and exhaust fan (see Figure 6.13). Powder is fluidized by an upward flow of conditioned air, while a binder solution is sprayed onto the suspended particles, forming liquid bridges that promote granule growth through collision and coalescence. Simultaneous drying removes solvent, stabilizing the granules. Process parameters control granule size and structure. Inlet air temperature typically ranges between 40 and 90°C. Superficial gas velocity is maintained just above the minimum fluidization velocity, ensuring a granulation time varying from minutes to tens of minutes.

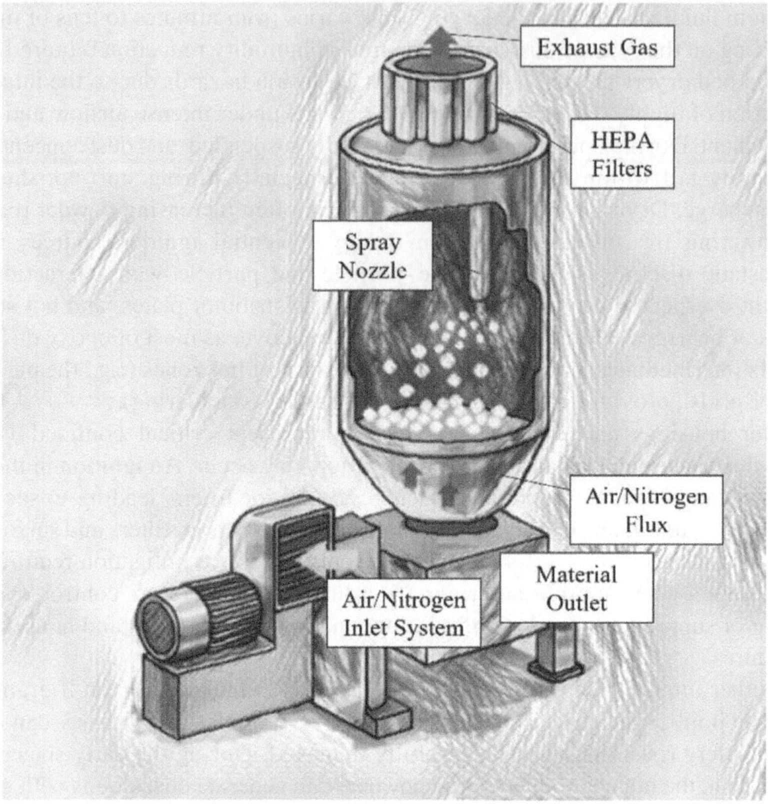

FIGURE 6.13 Schematic illustration of a fluid bed granulator for pharmaceutical industry.

As fluid bed driers, fluid bed granulators can exhibit a significant dust explosion hazard due to the intentional formation of fine, highly dispersed powder clouds under intense airflow, heating, and partial confinement. During the first phases of the granulation process, the combined hot fluidization and drying (which reduces moisture content and may lower the minimum ignition energy) can generate a dust concentration, which readily falls within the explosive range. Potential ignition sources include electrostatic discharges, hot inlet air, overheated heaters, and friction at distributor plates or filters. An internal ignition may propagate to filter housings or exhaust ducts, increasing explosion severity. Effective mitigation requires grounding, temperature control, inerting when necessary, explosion protection, and strict operational procedures.

6.1.5 Dust Collection Systems: Bag Filters and Cartridge Collectors

Dust collection systems are fundamental for dedusting of exhaust air from mills and dryers, product recovery from fluid bed granulators, and protection of downstream fans and environment. In pharmaceutical applications, mainly bag filters and cartridge collectors are used; therefore, they will be briefly discussed in the following.

A bag filter (or fabric filter) is a gas–solid separation device used to remove airborne dust and fine particulate matter from process exhaust air. It is also widely applied in milling, drying, granulation, and fluid-bed systems to ensure product recovery, operator protection, and environmental compliance. Its main components are: (a) a filter housing, that is an enclosed vessel, typically of AISI 304 or 316L stainless steel, suitably designed for either pressure or vacuum operations; (b) filter bags (sleeves) of different materials such as polyester, PTFE-coated, or antistatic media, suitably supported by internal cages; (c) gas inlet (dust-laden air) and outlet (clean air after filtration); (d) a cleaning system constituted by either a mechanical shaking or a reverse-pulse compressed air which dislodges accumulated dust cake; (e) a dust collection section formed by a hopper or collection bin with a discharge valve. Optional features include HEPA after-filters, explosion vents, isolation valves, and contained discharge systems (see Figure 6.14). Dust-laden air enters the housing and flows through the fabric filter bags. Particles are retained on the bag surface, forming a dust cake that enhances filtration efficiency. Clean air exits the filter, while periodic cleaning removes the dust cake, allowing it to fall into the collection hopper.

Bag filters represent one of the most critical units for dust explosion risk in pharmaceutical plants because they intentionally collect large quantities of fine, combustible dust under confined conditions. During normal operation, dust concentrations inside the housing and on filter surfaces can approach or exceed the minimum explosible concentration. Potential ignition sources include electrostatic discharges from filter media, hot particles carried from upstream equipment, and capable of igniting a filter media combustion, frictional heating, and faulty electrical components. Filter cleaning systems may resuspend dust clouds, increasing ignition likelihood. An internal ignition can result in a rapid pressure rise and flame propagation to connected process units. Effective mitigation requires antistatic filter media, grounding, explosion venting or suppression, isolation devices, and strict control of upstream ignition sources.

A cartridge collector is a high-efficiency dust collection system used to remove fine and ultrafine powders from process exhaust air. It performs the same function as a bag filter but uses pleated filter cartridges, which provide a large filtration area in a compact footprint. Cartridge collectors are widely applied where fine API dusts, highly reactive compounds, or space constraints are present. Its main components are: (a) a filter housing (typically of AISI 316L) designed for either vacuum or slight positive pressure; (b) a series of filter cartridges (pleated cylindrical elements) made of media such as cellulose-polyester blends, PTFE-coated, or antistatic materials; (c) a dust-laden air inlet which directs contaminated air toward the cartridge surfaces; (d) a clean air plenum and an outlet which collects filtered air for discharge or recirculation; (e) a cleaning system which uses reverse-pulse compressed air jets to dislodge dust cake; (f) a dust collection section constituted by a hopper (or bin) with contained discharge valve (see Figure 6.15). Dust-laden air enters the collector and passes through the filter cartridges from the outside inward. Particles are retained on the cartridge surface, forming a dust cake that enhances filtration efficiency. Periodic

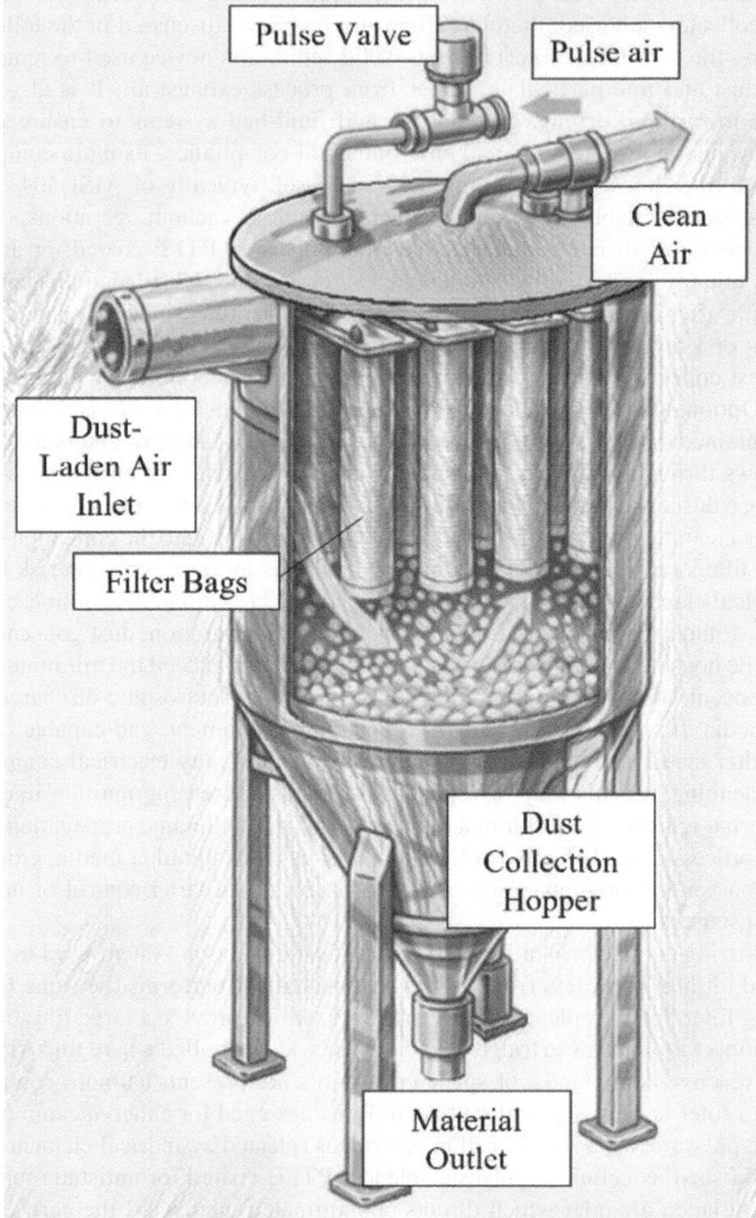

FIGURE 6.14 Schematic illustration of a bag filter for pharmaceutical applications.

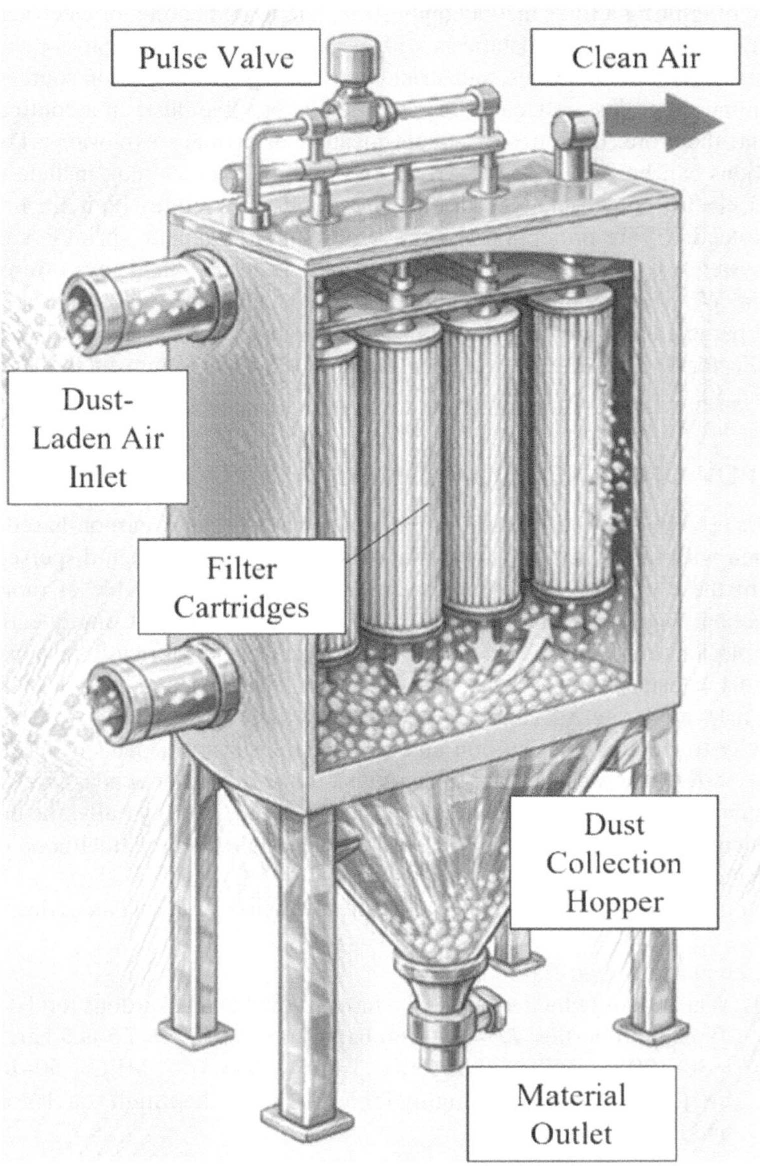

FIGURE 6.15 Schematic illustration of a cartridge collector for pharmaceutical applications.

pulse-jet cleaning removes the dust cake, allowing it to fall into the collection hopper, while clean air exits through the outlet.

Cartridge collectors present dust explosion hazards very similar to a bag filter. During operation and pulse-jet cleaning, dust layers are repeatedly re-suspended, potentially forming explosive dust clouds. Ignition sources may include electrostatic discharges from filter media, hot particles entering from upstream processes, and capable of igniting a filter media combustion, frictional heating, or electrical faults. Risk mitigation requires antistatic cartridges, proper grounding, explosion venting or suppression, isolation devices, and strict control of upstream ignition sources.

Summarizing, dust collectors concentrate combustible dusts in a confined environment; therefore, they are a frequent location of primary explosions. Dust concentrations can be far above the MEC, and ignition sources may include glowing embers, electrostatic discharges, or filter fires. Collectors require both fire and explosion protection. Fire protection typically consists of automatic sprinklers or water spray systems. Explosion protection may include venting, flameless venting, or suppression. Within this context, K_{st} and P_{max} values of the treated powder are fundamental parameters to be determined because they are essential for both vent sizing and collector design. Finally, explosion isolation is critical to prevent flame propagation back into process equipment.

6.2 POWDERS AND EQUIPMENT OF THE FOOD INDUSTRY

Powders used in the food industry are predominantly organic, carbon-based materials, often with high volatility upon thermal decomposition. When dispersed in air, many of these powders form explosive dust–air mixtures capable of rapid flame propagation, high pressure rises, and severe overpressures. Consequently, food dust explosions represent a significant industrial hazard, historically responsible for numerous catastrophic accidents. Food powders typically exhibit low MIE (often < 100 mJ), moderate K_{st} values (generally St 1 class), high P_{max} (7–10 bar), high sensitivity to electrostatic ignition and strong explosibility dependence on particle size (d_{50} < 100 μm is particularly hazardous), moisture (water acts as inertant), and fat content (which increases volatility and flame speed). Additionally, the degree of dispersion and turbulence is capable of significantly altering the likelihood of forming flammable dust clouds.

Food industry powders can be roughly divided into five main categories:

- Cereal and Starch-Based Powders
  1. Wheat flour (which is one of the most studied and hazardous food dusts). Typical properties: K_{st} = 80–160 bar·m/s (St 1), P_{max} = 7.5–9.5 bar, MIE = 30–100 mJ, MIT = 380–450°C, LIT = 250–300°C, MEC = 50–100 g/m^3. Its fine powders are highly ignitable due to their high starch content and large surface area.
  2. Corn starch (often more reactive than wheat flour). Typical properties: K_{st} = 120–200 bar·m/s, P_{max} = 8–10 bar, MIE = 20–50 mJ, MIT = 400–460°C, LIT = 250–280°C. Corn starch dust explosions are commonly used as reference cases in explosion testing laboratories.
- Sugar Powders

  Sucrose (granulated or powdered). Sugar dust explosions are well documented (e.g., sugar refinery accidents). Typical properties: K_{st} =

100–180 bar·m/s, P_{max} = 8–9 bar, MIE = 20–60 mJ, MIT = 370–450°C, LIT = 260–300°C, MEC = 30–60 g/m³. Fine powdered sugar is more hazardous than flour, primarily due to lower MIE.

- Milk and Dairy Powders
  1. Skim Milk Powder. Protein and lactose-rich dusts with significant explosion severity. Typical properties: K_{st} = 80–150 bar·m/s, P_{max} = 7–9 bar, MIE = 30–100 mJ, MIT = 390–430°C, LIT = 260–300°C.
  2. Whole Milk Powder. Fat content can increase not only flame propagation but also MIE values. Typical properties: K_{st} = 70–130 bar·m/s, P_{max} = 7–8.5 bar, MIE = 50–150 mJ.
- Cocoa, Coffee, and Beverage Powders
  1. Cocoa Powder. High fat content (10–25%) can lead to energetic explosions. Typical properties: K_{st} = 90–160 bar·m/s, P_{max} = 7.5–9 bar, MIE = 30–100 mJ, MIT = 380–420°C.
  2. Coffee Powder. Generally, it is less reactive than cocoa but still hazardous. Typical properties: K_{st} = 60–120 bar·m/s, P_{max} = 7–8 bar, MIE = 100–300 mJ.
- Protein and Additive Powders
  1. Soy Flour/Protein Isolates are widely used in food formulations. Typical properties: K_{st} = 90–180 bar·m/s, P_{max} = 8–9 bar, MIE = 10–50 mJ (very sensitive), MIT = 380–420°C. Proteins can decompose exothermically, contributing to high-pressure rise rates.
  2. Egg Powder. Highly reactive, especially when finely milled. Typical properties: K_{st} = 100–170 bar·m/s, P_{max} = 8–9 bar, MIE = 10–30 mJ.
- Spices and Additives

  Many spices are among the most sensitive food dusts. Examples are pepper, paprika, and garlic powder. Their typical properties are: K_{st} = 120–220 bar·m/s, P_{max} = 8–10 bar, MIE $<$ 10 mJ (extremely sensitive), MIT = 350–420°C. High oil content and very fine particle size explain their extreme ignitability.

Table 6.1 reports a comparative summary of the main explosive properties of dusts typically used in the food industry.

6.2.1 Raw Material Silos and Bins

Food industry silos are widely used for the storage of powdered raw materials such as flour, sugar, starches, milk powder, cocoa, and other finely divided food ingredients. These storage systems range from small metallic bins integrated into processing lines to large reinforced-concrete silos with capacities of thousand cubic meters. Silos are typically filled and emptied using pneumatic conveying systems or mechanical handling equipment, both of which promote dust generation and dispersion. Due to their geometry, silos inherently represent confined or semi-confined volumes, a

TABLE 6.1
Comparative Summary Table for Different Food Industry Dusts

Powder Type	K_{st} (bar·m/s)	P_{max} (bar)	MIE (mJ)	MIT (°C)
Wheat flour	80–160	7.5–9.5	30–100	380–450
Corn starch	120–200	8–10	20–50	400–460
Sugar	100–180	8–9	20–60	370–450
Milk powder	70–150	7–9	30–150	390–430
Cocoa	90–160	7.5–9	30–100	380–420
Soy protein	90–180	8–9	10–50	380–420
Spices	120–220	8–10	<10	350–420

condition that significantly increases the severity of potential dust explosions. As previously reported (see Table 6.1), numerous experimental studies and industrial accident investigations have shown that many food dusts exhibit low minimum ignition energy values, often below 50 mJ and, in some cases, below 10 mJ, making them highly sensitive to electrostatic discharges. The deflagration index of food powders typically falls within the St 1 class (0–200 bar·m/s), although finely milled starches, sugars, and protein powders may approach the lower end of the St 2 range under favorable conditions. The maximum explosion pressure is commonly in the range of 7–10 bar, which is sufficient to cause catastrophic structural failure in unprotected silos. During silo filling operations, especially when pneumatic conveying is employed, dust is readily dispersed, and airborne concentrations may exceed the minimum explosible concentration by several orders of magnitude. Such conditions are not limited to abnormal operation but may occur routinely during filling, discharge, or fluidization of powders. Furthermore, dust layers formed on internal surfaces can be re-entrained by pressure waves or air currents, contributing to secondary explosions with enhanced violence. Typical minimum ignition temperatures (MIT) for food dust clouds range from 350 to 450°C, while significantly lower values are observed for dust layers, increasing the risk of ignition from overheated surfaces or frictional heating.

Another relevant but often underestimated hazard in food industry silos is smoldering. Smoldering can be described as a slow, surface-driven oxidation reaction occurring within a porous solid at relatively low temperatures, typically 200–600°C, without visible flames. In food industry silos, smoldering develops inside bulk beds of combustible powders such as flour, starch, sugar, milk powder, cocoa, and protein-rich materials. Unlike gas-phase combustion, smoldering is diffusion-controlled by oxygen transport through interparticle voids. Moreover, it produces low heat release rates, allowing it to remain undetected for hours or days within large-stored masses, but it can generate hot spots capable of igniting dust clouds or layers. The porous structure and low thermal conductivity of food powders strongly favor smoldering propagation while limiting heat dissipation.

Smoldering is generally generated by localized ignition sources such as hot residues entering the silo from dryers or grinders, oxidation of fats and oils, or exothermic degradation of proteins or carbohydrates.

Given the high likelihood of dust dispersion and confinement, explosion protection is a critical design requirement for food industry silos. Explosion venting is the most commonly applied protective measure, and it is designed to limit internal overpressure to below the structural resistance of the silo. Vent sizing is based on experimentally determined K_{st} and P_{max} values, and it must account for silo volume, geometry, and dust reactivity. In many installations, venting is combined with damage-limiting or ducted venting concepts to control flame and pressure release in occupied or congested areas.

To prevent flame and pressure propagation to connected equipment, explosion isolation devices, such as fast-acting valves, rotary valves, or chemical isolation systems, are installed on pneumatic conveying lines and interconnecting ducts. In addition to passive and active protection systems, ignition source control remains essential. Grounding and bonding of all conductive components are mandatory to mitigate electrostatic discharge hazards, particularly given the low MIE of many food powders. Preventive measures also include temperature monitoring, avoidance of mechanical sparks, and strict housekeeping to limit dust accumulation.

6.2.2 Conveying Systems and Bucket Elevators

Conveying systems in food and grain processing industries include screw conveyors, belt conveyors, chain conveyors, pneumatic conveying lines, and bucket elevators. These systems are used to transport large quantities of combustible powders and granular materials between processing, storage, and packaging units. Among them, bucket elevators are particularly common in grain handling, milling, and food ingredient processing, due to their ability to move material vertically over significant heights with relatively compact footprints. However, their design and operating conditions make them one of the most critical items of equipment from a dust explosion safety perspective. In the following, screw conveyors, belt conveyors, and bucket elevators will be briefly described, discussing their criticalities in terms of dust explosion hazard.

A screw conveyor (or auger conveyor) is a mechanical conveying device used to transport powders, granules, or small solids over short to moderate distances. It is also used in pharmaceutical applications, where it commonly serves for controlled, enclosed transfer of APIs and excipients between processing steps such as milling, blending, drying, or packaging. Its main components are (a) a trough or tubular casing (U-shaped or fully enclosed tube) typically of AISI 316L stainless steel; (b) a screw (flight), that is a helical blade mounted on a central shaft; (c) a motor and a gearbox providing low to moderate rotational speed; (d) inlet and outlet ports for material feeding and discharge; (e) end bearings and seals (often isolated from product to meet GMP requirements). Optional features include variable-speed drives, heating/cooling jackets, and dust-tight or vacuum-rated designs (see Figure 6.16). As the screw rotates, material fed into the inlet is pushed forward by the rotating flights along the casing.

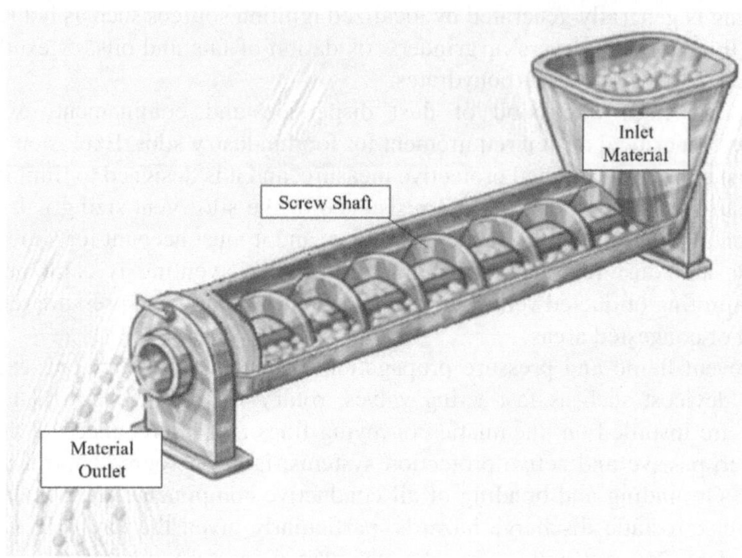

FIGURE 6.16 Schematic illustration of a screw conveyor for the food industry.

Conveying occurs primarily by drag and shear between the powder, the screw, and the casing. The system operates continuously, and it can provide metered, uniform feed rates. The typical length of the screw can be up to several meters, with a rotation speed spanning from low to moderate (to limit degradation).

Screw conveyors can present dust explosion hazards when handling fine, combustible powders due to the enclosed geometry and continuous mechanical motion. During operation, dust clouds may form at feed and discharge points, while deposits can accumulate inside the casing. Potential ignition sources include frictional heating between the screw flights and casing, mechanical sparks caused by foreign objects or misalignment, overheated bearings, and electrostatic discharges from poorly grounded components. The confined trough can allow pressure buildup if ignition occurs, with possible flame propagation to connected equipment. Effective risk mitigation requires proper grounding, clearance control, temperature monitoring, explosion isolation, and careful design of inlet and outlet interfaces.

A belt conveyor is a continuous conveying system used to transport bulk materials, powders, or packaged goods over short to long distances. It consists of a moving flexible belt supported on rollers or a flat bed and driven by a motorized pulley. Belt conveyors are widely used where gentle handling and high throughput are required. Its main components are: (a) a conveyor belt made of rubber, polymer, fabric-reinforced, or metal, selected for chemical compatibility with the transported powder and wear resistance; (b) a drive pulley powered by an electric motor and a gearbox; (c) a tail pulley which provides belt tension and return guidance; (d) idlers or a support bed capable of support the belt along its length; (e) feed and discharge zones (chutes or hoppers for material transfer). Optional features include covers, dust skirts, scrapers, and variable-speed drives (see Figure 6.17). Material is loaded onto the belt at the feed point. As the belt moves, friction between the belt and drive pulley transports the material along the conveyor until it is discharged at the outlet by gravity or a deflector.

Belt conveyors present dust explosion hazards primarily at loading, transfer, and discharge points, where fine combustible powders can become airborne and form explosive dust clouds. Unlike enclosed conveyors, belt systems often allow dust dispersion into surrounding areas, increasing the risk of secondary explosions from accumulated deposits. Potential ignition sources include frictional heating from belt misalignment, slipping on drive pulleys, overheated bearings, electrostatic discharges from non-conductive belts, and electrical faults. Dust layers on structures, rollers, and floors can be re-suspended by air currents or disturbances. Effective risk mitigation requires dust containment, antistatic belts, proper grounding, belt tracking control, temperature monitoring, housekeeping, and explosion isolation where conveyors connect to enclosed process equipment.

A bucket elevator is a vertical conveying system used to transport bulk solids, powders, or granules upward between different process levels. Material is carried in a series of buckets attached to a moving belt or chain, making bucket elevators essential where space is limited and vertical lifting is required. Its main components are: (a) a boot section composed of a lower housing where material is fed and containing the take-up pulley (or sprocket); (b) buckets made of either metal or polymer, mounted at regular intervals on a belt or chain; (c) a belt (or chain) which provides bucket movement (it is selected based on load and temperature); (d) a head section

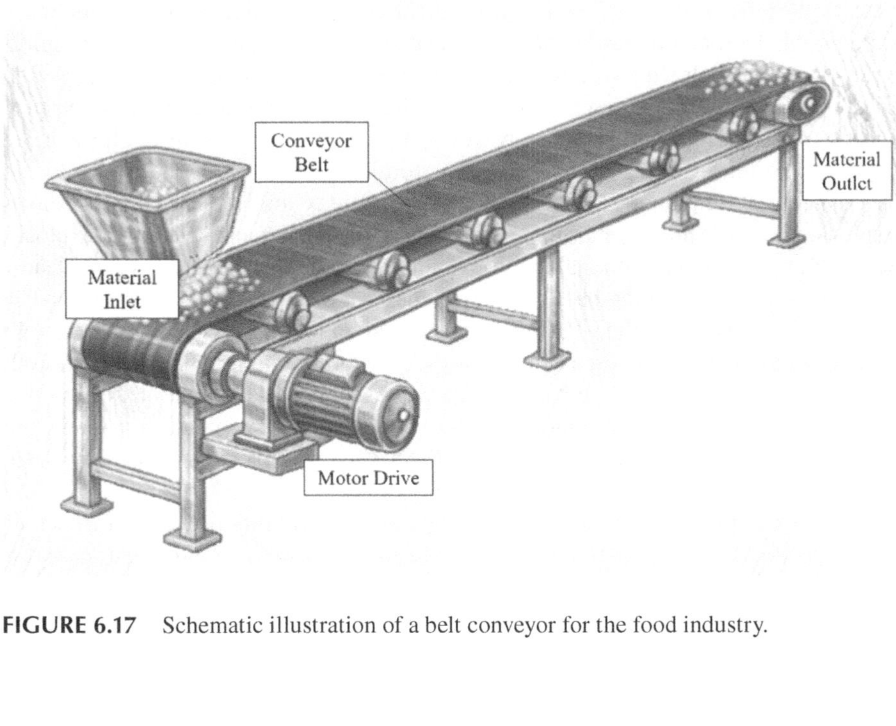

FIGURE 6.17 Schematic illustration of a belt conveyor for the food industry.

composed of a upper housing with drive pulley or sprocket for material discharge by centrifugal force or gravity; (e) a casing (trunk) composed by an enclosed vertical column guiding bucket motion; (f) a motor and a gearbox located at the head section (see Figure 6.18). Material is introduced at the boot, where buckets scoop it up. As the belt or chain moves upward, the filled buckets carry material to the head section, where it is discharged into a chute by centrifugal action or gravity, depending on elevator type.

Bucket elevators are among the most critical equipments for dust explosion hazards due to the combination of fine combustible dust, high mechanical energy, and often strong confinement. Dust clouds can form in the boot section during loading and in the head section during discharge, often reaching explosive concentrations. Potential ignition sources include belt or chain friction, misalignment, overheated bearings, metal–metal contact, and entrained hot particles. The vertical casing promotes flame acceleration and pressure piling, allowing an initial ignition to propagate rapidly and violently. Accumulated dust deposits further increase the risk of secondary explosions. Effective mitigation requires ignition source control, monitoring of belt speed and temperature, explosion venting or suppression, isolation devices, and rigorous housekeeping.

Summarizing, most conveying equipment operates in a fully or partially enclosed configuration, providing the confinement necessary for pressure build-up in the event of a dust explosion. During normal operation, dust clouds are generated by material transfer, bucket filling and emptying, belt movement, and leakage points. In bucket elevators, dust concentrations can locally exceed the minimum explosible concentration, particularly in the boot section and head casing, even during routine operation. Bucket elevators have historically been involved in some of the most severe dust explosions in the food and grain industries. This is due to a combination of unfavorable factors: high airborne dust concentrations, frequent mechanical ignition sources such as belt slippage, overheated bearings, and frictional heating, as well as the vertical geometry, which promotes rapid flame acceleration and pressure piling. Once ignited, flames and pressure waves can propagate upward through the elevator casing, often resulting in extensive structural damage and secondary explosions in connected equipment. The dust deflagration index and the maximum reduced explosion pressure are key parameters governing the design and spacing of explosion vents or suppression systems along conveying equipment. Higher K_{st} values lead to faster pressure rises, reducing the effectiveness of venting if not properly distributed. At the same time, the MEC provides an indication of how easily an explosible atmosphere may form during normal or upset operating conditions, highlighting the importance of dust-tight design and effective housekeeping. In situations where venting is impractical, such as indoor installations, explosion suppression systems may be applied to rapidly detect and extinguish an incipient explosion before destructive pressures develop.

Effective explosion isolation at both inlet and outlet points is essential to prevent flame and pressure propagation into upstream and downstream equipment, including silos, filters, and conveying lines. In addition to protection systems, preventive measures play a critical role in risk reduction. Continuous bearing temperature

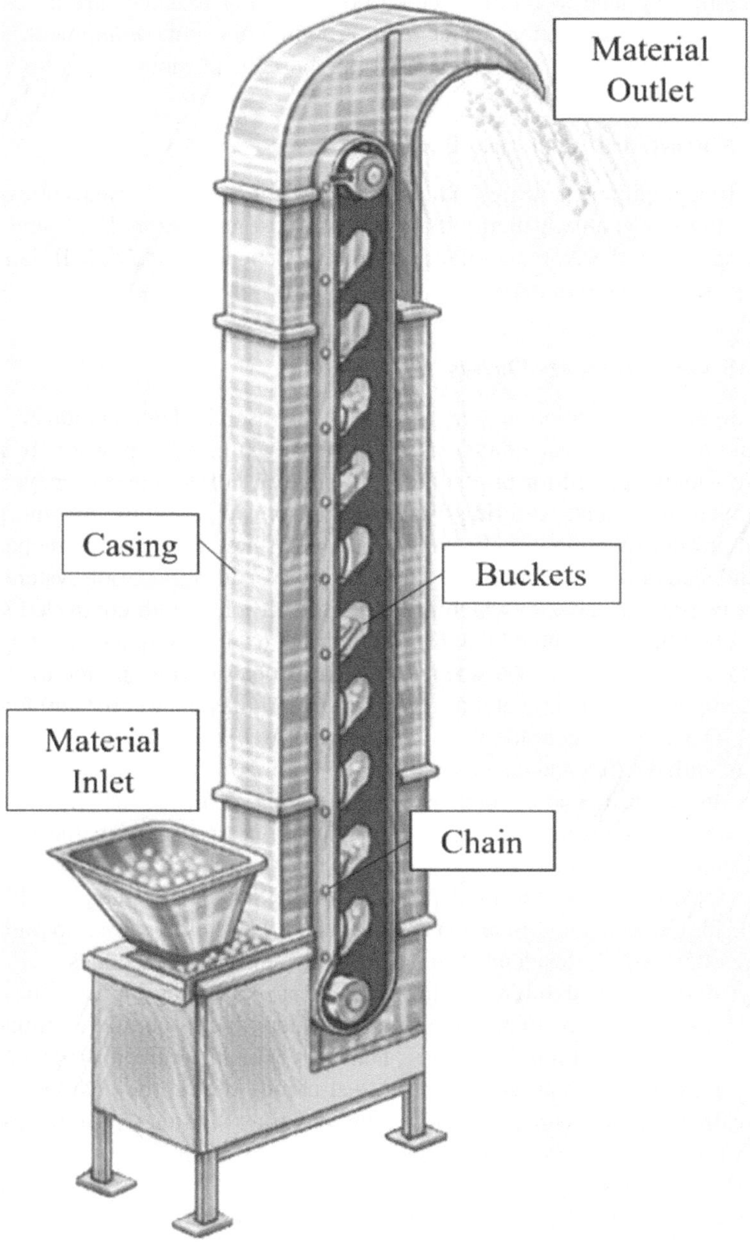

FIGURE 6.18 Schematic illustration of a bucket elevator for the food industry.

monitoring, belt alignment sensors, and speed monitoring help identify mechanical faults before they lead to ignition. Good housekeeping practices are necessary to minimize dust accumulation inside and around conveying equipment, thereby reducing both the likelihood and severity of potential explosions.

6.2.3 Milling and Grinding Equipment

Food milling equipment shares many characteristics with pharmaceutical mills, including high mechanical energy input and fine dust generation. Explosion protection strategies are, therefore, similar and strongly dependent on K_{st}, MIE, and MEC values of the processed material.

6.2.4 Spray and Freeze Dryers

A spray dryer is a continuous drying process widely used in the food industry to convert liquid feeds (solutions, emulsions, or suspensions) into dry powders in a single step. It is widely applied for producing milk powder, coffee, flavors, enzymes, and nutraceutical ingredients, offering excellent control over moisture content, particle size, and powder functionality. Its main components are: (a) feed system constituted by a liquid feed tank; (b) a pump and a flow control; (c) an atomization system (rotary atomizer or pressure nozzle) which produces fine droplets with controlled size; (d) a drying chamber constituted by a large cylindrical (or less frequently, conical) vessel which provides sufficient residence time for droplet drying; (e) an air handling system composed of an air heater (direct or indirect), and a blower (or fan) for air circulation; (f) a powder separation system (cyclones and/or bag filters); (g) an exhaust air system with ducting and an exhaust fan (see Figure 6.19). The liquid feed is atomized into fine droplets and contacted with hot drying air inside the chamber. Rapid evaporation occurs, forming dry particles that are separated from the air stream and collected, while exhaust air is discharged. Inlet air temperature typically ranges between 150 and 220°C for food products (120-180°C for proteins, APIs, etc..; 70-120°C for thermally sensitive pharma products), while outlet air temperature is in the range of 50–90°C. Residence time is very low: up to a few seconds.

Spray dryers can exhibit a significant dust explosion hazard due to the simultaneous presence of fine, combustible powders, high operating temperatures, and large, partially confined volumes. During drying, atomized droplets rapidly form dry particles that can create dense dust clouds within the chamber, and the downstream equipment (such as cyclones and bag filters), often within the explosive concentration range. Potential ignition sources include hot drying air, overheated air heaters, deposits igniting on hot surfaces, electrostatic discharges, and sparks from mechanical or electrical equipment. Powder deposits inside ducts and downstream collectors increase the risk of secondary explosions. An initial ignition can propagate through the exhaust system with severe consequences. Effective mitigation requires inert gas operation when feasible, temperature control, grounding, explosion venting or suppression, explosion isolation, and rigorous housekeeping.

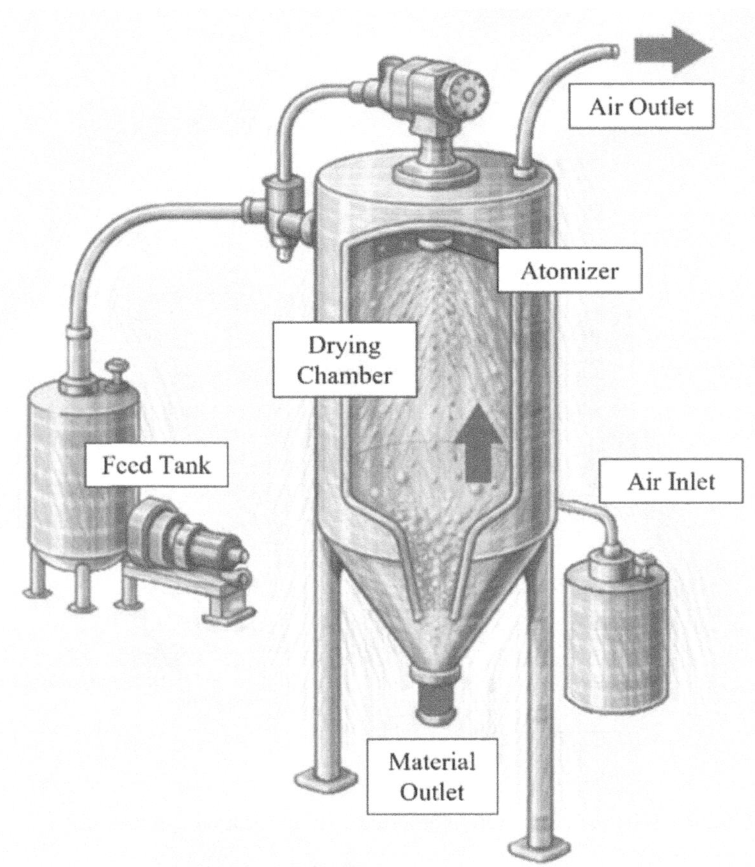

FIGURE 6.19 Schematic illustration of a spray dryer for food applications.

A freeze dryer, also known as a lyophilizer, is a batch drying system used to remove water or solvents from products by sublimation, that is, the direct transition of ice from solid to vapor under low temperature and reduced pressure. This process preserves product structure, stability, and biological activity, making it ideal for heat-sensitive materials. Its most common applications concern pharmaceutical injectables and vaccines, biologics and enzymes, food ingredients and nutraceuticals. Its main components are: (a) a drying chamber, that is a vacuum-tight vessel, typically of stainless steel, which contains shelves to hold product trays (or vials); (b) shelves (heat transfer plates) actively temperature-controlled (cooling and heating); (c) a refrigeration system which freezes the product before drying; (d) a vacuum system composed of a vacuum pump and valves to maintain low pressure; (e) a condenser (cold trap) which captures sublimated vapor as ice; (f) a control system which regulates temperature, pressure, and cycle steps (see Figure 6.20). The product is first frozen on the shelves. Under vacuum, gentle heat is applied, causing ice to sublime

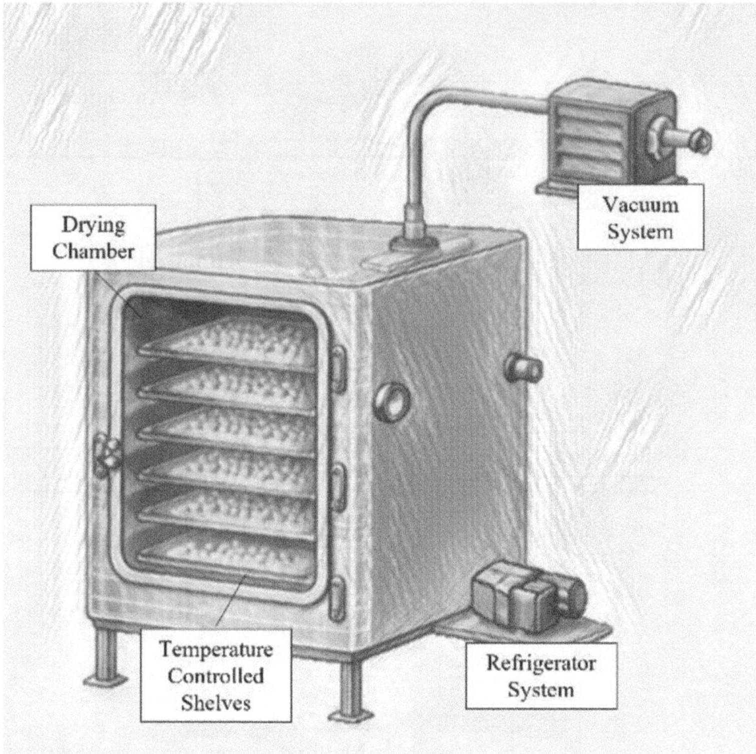

FIGURE 6.20 Schematic illustration of a freeze dryer for food applications.

during primary drying. Residual bound moisture is removed during secondary drying at higher temperatures. Vapor is captured on the condenser.

Freeze dryers generally present a lower dust explosion risk than high-energy drying technologies; however, hazards may arise when processing dry, combustible powders, particularly during unloading, milling after drying, or product handling inside the chamber. During secondary drying, very low moisture content can increase powder reactivity and reduce the minimum ignition energy. Potential ignition sources include electrostatic discharges under low-pressure conditions, electrical faults, and hot surfaces associated with shelf heating. If air ingress occurs, a combustible dust–air atmosphere may form within the confined chamber. Accumulated dust deposits can also contribute to secondary explosions. Effective mitigation relies on inerting or vacuum integrity, grounding, controlled discharge procedures, and careful management of downstream powder handling operations.

Summarizing, both spray and freeze dryers can exhibit dust explosion hazards, especially when the low-moisture-content final powder is formed. Explosion venting is applied to the dryer chamber, and to all downstream equipment (such as cyclones and dust collectors). Isolation between connected units is essential, even for fire protection.

6.2.5 Cyclones and Electrostatic Precipitators

Cyclones and electrostatic precipitators are used for different gas–solid separations in either food or pharmaceutical industries (from product recovery to dust-laden air purification purposes).

A cyclone is a mechanical gas–solid separation device used in food processing to remove solid particles (powders, fines, or droplets) from an air or gas stream using centrifugal forces. It contains no moving parts, and it is widely applied for product recovery (e.g., from spray dryers), air cleaning (dedusting of dryers and mills), protection of downstream equipment, and separation of flour, milk powder, sugar, and starch fines. Its main components are: (a) an inlet duct (tangential or scroll-type entry) for particle-laden air; (b) a cylindrical body which initiates the swirling (vortex) motion; (c) a conical section which is mainly devoted to either address or collect the separated particles; (d) a vortex finder (outlet tube) which removes cleaned air from the top; (e) a dust discharge outlet, located at the cone apex, often with an airlock (see Figure 6.21). All product-contact surfaces are typically stainless steel and designed for hygienic operation. Particle-laden air enters tangentially, creating

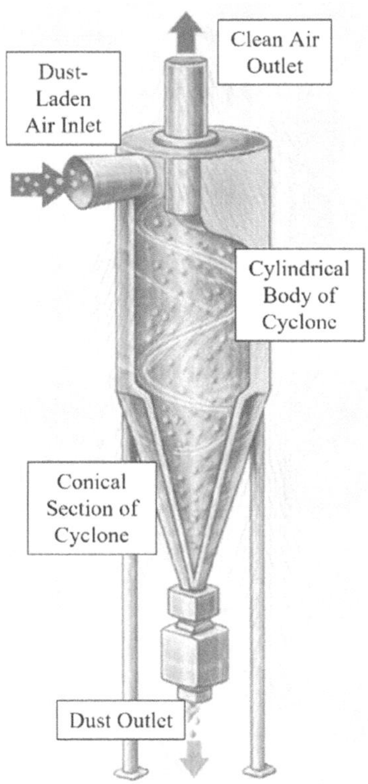

FIGURE 6.21 Schematic illustration of a cyclone for food industry applications.

a rotating vortex. Centrifugal forces drive heavier particles toward the wall, where they lose momentum and fall into the dust outlet, while cleaned air exits through the vortex finder.

Cyclones present dust explosion hazards because they handle airborne combustible powders under turbulent flow and partial confinement. During operation, dust concentrations inside the cyclone can approach or exceed the minimum explosible concentration, especially downstream of dryers, mills, or spray dryers. Although cyclones have no moving parts, potential ignition sources include electrostatic discharges from particle–wall interactions, hot particles entrained from upstream

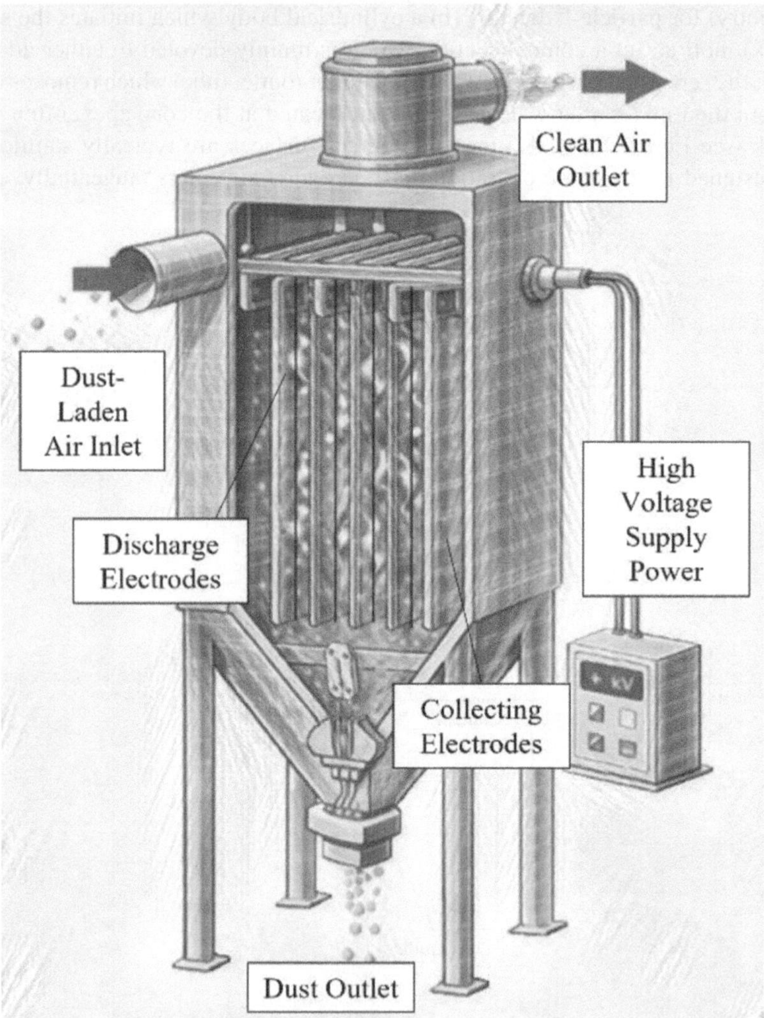

FIGURE 6.22 Schematic illustration of an electrostatic precipitator for food applications.

processes, and high surface temperatures. The swirling flow can promote flame acceleration if ignition occurs, and an explosion may propagate to connected ducts, filters, or dryers. Accumulated dust in the cone or discharge device increases the risk of secondary explosions. Effective mitigation requires grounding, temperature control, explosion venting, isolation, and proper discharge sealing.

An electrostatic precipitator (ESP) is a gas–solid separation device used to remove fine particulate matter, aerosols, or dust from process exhaust gases by means of electrostatic forces. In the food industry, ESPs are applied where very fine particles must be captured with high efficiency and low pressure drop, often downstream of dryers, roasters, or thermal processes. Its main components are: (a) an ESP housing, that is an enclosed metal casing, often of stainless steel for hygienic applications; (b) discharge electrodes, that is high-voltage wires or needles that ionize the gas; (c) collecting electrodes (plates or tubes), that is grounded surfaces where charged particles are deposited; (d) a high-voltage power supply which generates a strong electric field (typically several kV); (e) a dust removal system (using either a rapping mechanism, or a washing system, or a gravity discharge) (see Figure 6.22). The working principle is quite simple. Process gas flows through the ESP and passes the discharge electrodes, where particles become electrically charged. Under the electric field, charged particles migrate toward the collecting electrodes and adhere to them. Periodic cleaning removes the collected material, which is discharged, while cleaned gas exits the unit.

Electrostatic precipitators present specific dust explosion hazards because they intentionally combine fine combustible dust, strong electric fields, and confined process volumes. During operation, high dust loadings may be present on collecting electrodes and in hoppers, while rapping or cleaning can resuspend dust clouds. The high-voltage system itself represents a potential ignition source through electrical discharges, sparks, or arcing, especially if dust layers accumulate on electrodes. Hot particles entering from upstream processes can further increase ignition likelihood. An internal ignition may propagate to connected ducts or downstream equipment. Effective risk mitigation requires careful control of dust properties, grounding, explosion venting or suppression, isolation devices, temperature monitoring, and strict electrical safety management.

Summarizing, cyclones handling combustible dust require explosion venting unless self-venting criteria are met. Electrostatic precipitators must not operate above the MEC due to ignition risk from arcing.

7 Safety Applications in the Food and Pharma Industries

7.1 QUANTITATIVE RISK ASSESSMENT METHODS: BRIEF OVERVIEW

Over recent decades, Quantitative Risk Analysis (QRA) [1–4] has become firmly established as a fundamental approach for conducting risk assessments, primarily because it enables the association of an objective measure, such as probability, with an undesired event. Nevertheless, QRA represents a broad conceptual framework, and it does not, by itself, prescribe a specific methodological pathway or analytical procedure. In practice, a QRA is generally articulated through multiple stages, including the identification of potential accident scenarios and the evaluation of both the severity and likelihood of unwanted events [5]. Over time, several methodologies have been proposed to enhance the effectiveness of QRA. With respect to the identification of accidental scenarios, the Hazard and Operability (HazOp) technique, originally introduced by ICI (Kletz) in the early 1960s [6], was developed to classify process-related criticalities in the chemical industry by systematically examining deviations of process variables. At present, HazOp remains one of the most widely applied hazard identification techniques within chemical companies. The strength of the HazOp method lies in its ability to identify hazardous scenarios and critical plant sections, and its structured approach allows for the generation of objective and reproducible results. However, the outcomes of a HazOp study consist essentially of unstructured lists of causes and consequences associated with process variable deviations. Consequently, HazOp is not well-suited for estimating accident probabilities, which represent a key element in the definition of a quantitative risk index.

To overcome these limitations, other methods have been developed with the objective of improving risk estimation. Failure Mode and Effects Analysis (FMEA), as well as Failure Mode, Effects and Criticality Analysis (FMECA) [5, 7], are techniques specifically aimed at identifying accident scenarios based on the failure modes of individual components within process equipment, such as valves, switches, and other mechanical or electrical devices. In addition to the classical FMEA framework, FMECA introduces a criticality index that allows the analyst to focus attention on the most critical components of the plant. As a result,

DOI: 10.1201/9781003645320-7

FMEA and FMECA represent effective tools for identifying components involved in potential accidents, and the corresponding failure rates can often be retrieved from available literature databases. Nonetheless, these techniques also yield largely unstructured information, and they do not directly provide an estimation of the probability associated with a given accidental scenario. For probability estimation, which is essential for quantitative risk assessment, Fault Tree Analysis (FTA) [8, 9] is among the most adopted methodologies, owing to its high degree of flexibility across different application domains. FTA is based on the logical combination of basic events through logic gates; however, its implementation can become particularly complex, especially in large and intricate plants characterized by numerous interactions between human operators and technical systems. It, therefore, becomes evident that an effective risk assessment can only be achieved through the integration of multiple complementary methods, each addressing specific phases of the risk analysis process.

Within this context, several integrated methodologies have been extensively developed and proposed, including MAXCRED [10] and SCAP [11]. These approaches are designed to identify credible accident scenarios in chemical process industries, such as fires and explosions, while simultaneously evaluating the magnitude of accident consequences through simulation techniques and defining individual, plant-wide, and societal risk indicators. MAXCRED is primarily focused on the assessment of accident consequences, whereas SCAP is oriented toward safety management, and it is applied to define and evaluate the effectiveness of risk mitigation measures. In addition, dynamic modeling approaches have been proposed, such as the bow-tie methodology introduced by Khakzad et al. [12], which are particularly useful for continuously updating probability estimates of Top Events (TE) in highly dynamic systems. A significant limitation common to many of these approaches is their limited generality in representing the underlying mechanisms of industrial accidents. Because they aim to cover a broad range of chemical processes, their effective application often requires a highly skilled analysis team capable of identifying critical process features and translating them into a coherent Fault Tree (FT) structure. Historically, an initial attempt to directly integrate HazOp results with FTA was represented by the development of Recursive Operability Analysis (ROA), proposed by Piccinini and Ciarambino [13, 14]. The core idea of ROA is to focus on process variable deviations, following a HazOp-like structure, but to interconnect these deviations through cause–sequence relationships rather than merely listing them. ROA has been successfully applied to risk assessments in the chemical industry [15, 16], and further refinements have been introduced over time. Despite these improvements, ROA exhibits certain drawbacks: it does not explicitly account for plant states in which protection systems operate correctly, and it can be time-consuming, requiring the compilation of numerous records even for relatively simple systems [17]. More recently, the original ROA methodology has been updated through the introduction of a Cause Consequence Diagrams (CCD) approach [17, 18]. This technique is based on a highly structured ROA table, closely resembling a traditional HazOp analysis, from which a complete fault tree can be systematically derived. The method allows the analyst to consider both scenarios involving the failure of automatic and manual protection systems and scenarios in which such protections function as intended. Compared to the original ROA formulation, ROA-CCD can be applied more efficiently, as demonstrated by Contini et al. [17]. However, because ROA-CCD remains fundamentally rooted in a HazOp-based framework, it does not explicitly indicate which process variable deviations are most relevant for the occurrence of accidental TE. Successive works developed a method that retained the structural characteristics of ROA-CCD while enabling the almost automatic generation and solution of fault trees starting from basic process information. Given the wide variety of possible industrial accidents, it is acknowledged that no single method can be universally applicable to all chemical plants. Among all approaches, the Recursive Operability Analysis Plus with Incidental Sequence Diagrams (ROA Plus–ISD) has been specifically designed to identify accidental scenarios and estimate their associated probabilities, with particular emphasis on dust explosion hazards. Dust explosions represent extremely severe and often fatal accidents in process industries [19], and the development of reliable methodologies for assessing the risks associated with combustible dusts has, therefore, attracted significant interest [20]. When considering process variables in relation to dust explosions, a direct correspondence can be established with the well-known dust explosion pentagon [21]. According to this model, five conditions must be simultaneously satisfied for a dust explosion to occur: the presence of combustible dust, the availability of an oxidizing agent (typically oxygen), the dispersion of dust in air, confinement, and the presence of an ignition source.

The ROA Plus–ISD methodology is structured around the a priori definition of all relevant process variables, human operations, and plant components, organized according to the conditions identified by the dust explosion pentagon. The basic events required for fault tree computation are derived through the execution of a FMEA. Once all necessary elements have been identified, the ROA table can be constructed, considering exclusively those deviations that are relevant to dust explosion scenarios. Subsequently, for each individual record of the ROA table, the corresponding Incidental Sequence Diagram is generated. By iterating this procedure for all records, the final fault trees associated with the identified Top Events are automatically obtained through the logical matching of correlated ISDs. The resulting fault trees can then be analyzed using dedicated software tools, such as PROFAT [22], OpenFTA, FaultTree+ [9], or ASTRA [23].

The principal distinction between ROA Plus–ISD and ROA-CCD methodologies lies in the renewed adoption of Incidental Sequence Diagrams for fault tree generation, while preserving the overall structure of the ROA-CCD table. The flowchart illustrating the ROA Plus–ISD methodology is shown in Figure 7.1. The overall procedure may be succinctly described through the following sequence of steps: (a) identification of combustible materials and oxidizing agents; (b) subdivision of the plant into functional nodes; (c) inventory of plant components and execution of a Failure Mode and Effects Analysis; (d) identification of human interactions with the system, together with the associated potential human errors; (e) definition of the process variables and possible ignition sources relevant to dust explosion phenomena;

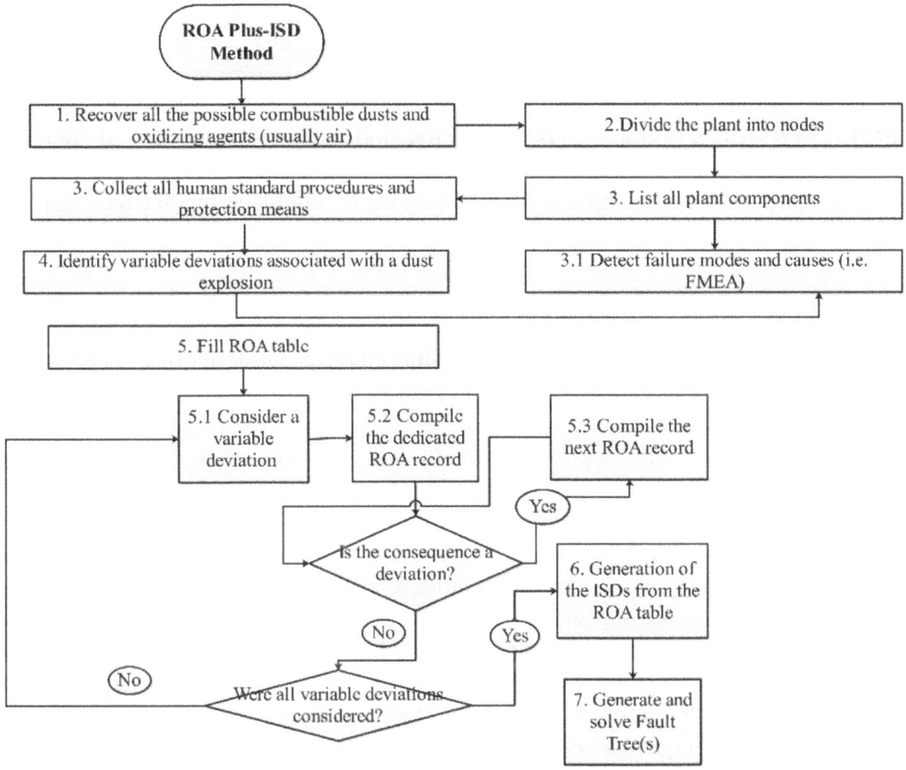

FIGURE 7.1 Flowchart of the ROA Plus–ISD method for dust explosions.

(f) compilation of the ROA Plus table; (g) generation of the corresponding Incidental Sequence Diagrams and related Fault Trees; and, finally, execution of the Fault Tree Analysis.

Regarding the preliminary phase of the methodology, the technical aspects are described in detail. The initial step consists of identifying all substances present within the plant, including combustible dusts and oxidizing agents. At this stage, it is essential to collect all relevant information concerning dust explosibility [24]; therefore, parameters such as Minimum Explosive Concentration (MEC), Minimum Ignition Energy (MIE), Limiting Oxygen Concentration (LOC) [25], Minimum Ignition Temperature (MIT) [21], Layer Ignition Temperature (LIT), and the deflagration index (K_{St}) [26] must be known.

If it is possible for dust concentrations and/or other characteristic parameters to reach critical values within the plant under any operating condition, then a dedicated dust explosion risk analysis becomes necessary. Subsequently, the plant must be subdivided into nodes, ideally distinguishing between different functions performed by the available equipment. In order to define the basic events that may initiate accidental scenarios, all relevant components, such as switches, safety systems, elevators, conveyor belts, mixers, collectors, and valves, must be identified and listed. When available, the Piping and Instrumentation Diagram (P&ID) represents the most suitable reference for this task. Based on this inventory, failure modes and their associated effects must be determined. In this context, the application of FMEA or FMECA [5] represents an effective and strongly recommended approach for the correct implementation of the method. FMEA is carried out by completing the standard scheme reported in Table 7.1, and it should be emphasized that the identified failure causes constitute the basic events that will appear in the final Fault Trees.

In addition to equipment-related failures, procedures involving human operators must also be identified, as they may play a critical role in safety-related issues [27]. This includes both routine operational activities and protective actions, such as manual loading operations or the opening of hand-operated valves. Human actions and associated errors can be analyzed through a FMEA-like framework, in which each interaction between the operator and the system is examined to identify potential errors. The probabilities of such errors can be retrieved from established literature databases [5, 28].

The subsequent step involves the definition of the process variables relevant to dust explosion phenomena. These deviations, derived from the well-known dust explosion pentagon, are summarized in Table 7.2. It is important to underline that these process variables must be explicitly associated with specific items of equipment and with the substances involved. The variables should be defined in a form suitable for inclusion within the ROA table, thereby allowing the systematic generation of Incidental Sequence Diagrams.

Figure 7.2 presents either the most basic Fault Tree or the modified ROA-ISD representation of a dust explosion.

By adopting one of these two representations, the presence of dust within the explosible range can be directly described through process variables. Additional variables listed in Table 7.2, such as flow, mass, and pressure, can contribute to the generation of airborne dust; however, since these are standard process variables, the ROA framework is already well-suited to address their interactions. Elevated mass and flow rates may lead to the formation of deposits, which, in turn, can promote the generation of dust clouds.

Dispersion in air may arise from multiple mechanisms, including primary explosions, equipment specifically designed for dust transport, or interactions with

TABLE 7.1
FMEA Table for ROA Applications

Component	Description of the Function	Failure Modes	Failure Causes	Consequences	Protections		Frequencies
					Alarms (Optical/ Acoustic)	Automatic Protective Means	

TABLE 7.2
Process Variables Related to the Explosion Pentagon

Condition	Variable Deviation	Symbol
Presence of combustible dust	High concentration (>MEC)	$x.h.C_i$
	High mass	$x.h.m_i$
	High Flow	$x.h.F_i$
Presence of oxidizing agent	High concentration (>LOC)	$x.h.C_i$
Ignition Source	High Metallic Surface Temperature (>LIT)	$x.h.T_m$
	High Temperature (>MIT)	$x.h.T$
	High Energy Spark (>MIE)	$x.h.E$
	High Enthalpy (>MIE)	$x.h.\Delta H$
Dispersion	High Air/Nitrogen Flow	$x.h.F_{air/N2}$
Confinement	Low Volume	$x.l.V$
	High Pressure	$x.h.P$

x indicates a generic node, i indicates the substance which the variable is referred to.

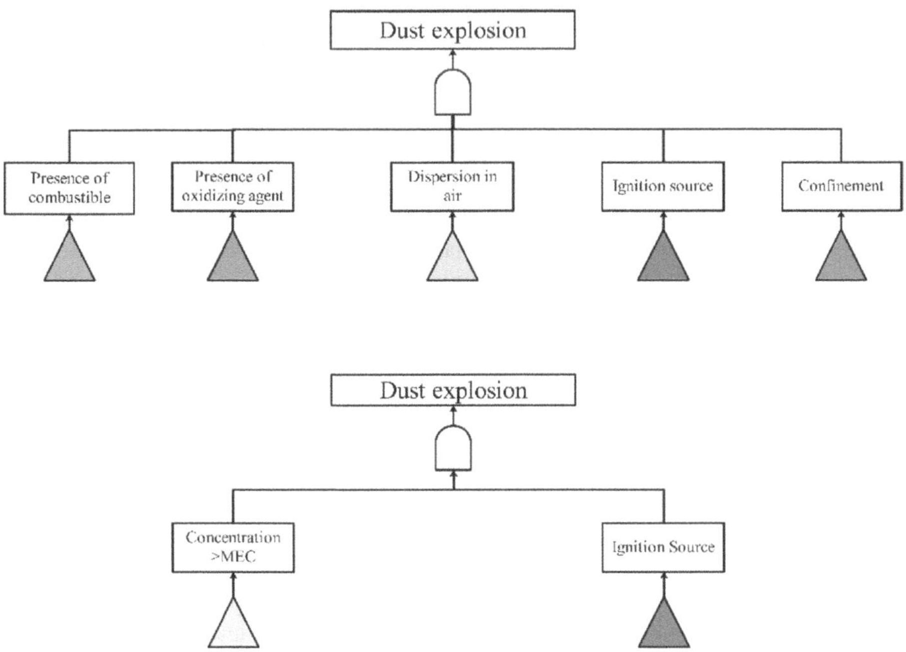

FIGURE 7.2 Classical (a) and modified (b) dust explosion fault trees (both representations can be used).

airflows [21], and these phenomena are typically represented by high pressure or high flow conditions.

Confinement is generally associated with the presence of enclosures within the plant, corresponding to a reduced available volume. Although this variable is not usually monitored during operation, it becomes particularly relevant during the design of new plant sections. In practice, low available volume often behaves as a Boolean variable, indicating whether confinement is present or absent within the facility.

The identification of ignition sources represents an additional challenge, as numerous potential sources exist, including electrical sparks, friction sparks, brush discharges, fires, and hot surfaces capable of exposing dust clouds to temperatures exceeding the MIT [21]. Most of these ignition mechanisms can be described as deviations of either standard or non-standard process variables (or functions). In the first group of variables, it is possible to find high temperature (of either surfaces or dust clouds); otherwise, in the other group of variables/functions it is possible to find energy in all its forms (e.g., sparks, enthalpy from exothermic reactions, etc.). Especially for non-standard process variables/functions, all potential ignition sources must be carefully identified and listed a priori before initiating the compilation of the ROA table.

Once the preliminary steps have been completed, the central phase of the methodology consists in performing the Recursive Operability Analysis (Plus), which involves compiling the table structured according to the standard format reported in Table 7.3.

Table 7.3 corresponds to the classical ROA-CCD formulation [17, 18], which represents a more articulated evolution of the original ROA table [13, 16, 29]. In particular, Column 1 (referred to as the Node Deviation Variable, NDV) effectively merges the three elements of the traditional HazOp framework into a single descriptor, encompassing the plant node, the relevant physical variable, and the type of deviation. Column 2 is devoted to identifying the causes of the deviation, whereas Column 3 describes the resulting consequences on the plant, assuming the failure of all protective measures. These consequences may consist either of further deviations of process variables, such as an increase in pressure following a temperature rise, or of a Top Event, for example, a fire. The table explicitly considers the presence of protection systems and distinguishes between alarm functions, which typically require operator intervention, and automatic protective devices. When either causes or consequences manifest themselves as deviations of additional process variables, the analysis must be iteratively extended until only basic events and TEs remain. At this stage, the principal distinction between the ROA-CCD approach and ROA-ISD methodology becomes evident. Within the ROA-CCD framework, once the ROA table has been fully completed, a Cause–Consequence Diagram can be derived from each individual record (REC), that is, from each row of the table, as illustrated in Figure 7.3.

This diagram explicitly includes information regarding the plant condition when protective systems operate correctly, thereby enabling a more detailed and accurate interpretation of the system behavior. Nevertheless, Cause–Consequence Diagrams

TABLE 7.3
ROA-CCD Scheme

					Protections				
					Manual				
REC	NDV	Causes	Consequences due to the Protections Failures	Plant State with Protections Working Correctly	Alarms (Optical/ Acoustic)	Operator Actions on Component	Automatic Protective Systems	Frequencies	TE
	1	2	3	4	5	6	7		

(Legend: REC→Record, NDV→Node Deviation Variable, TE→Top Event).

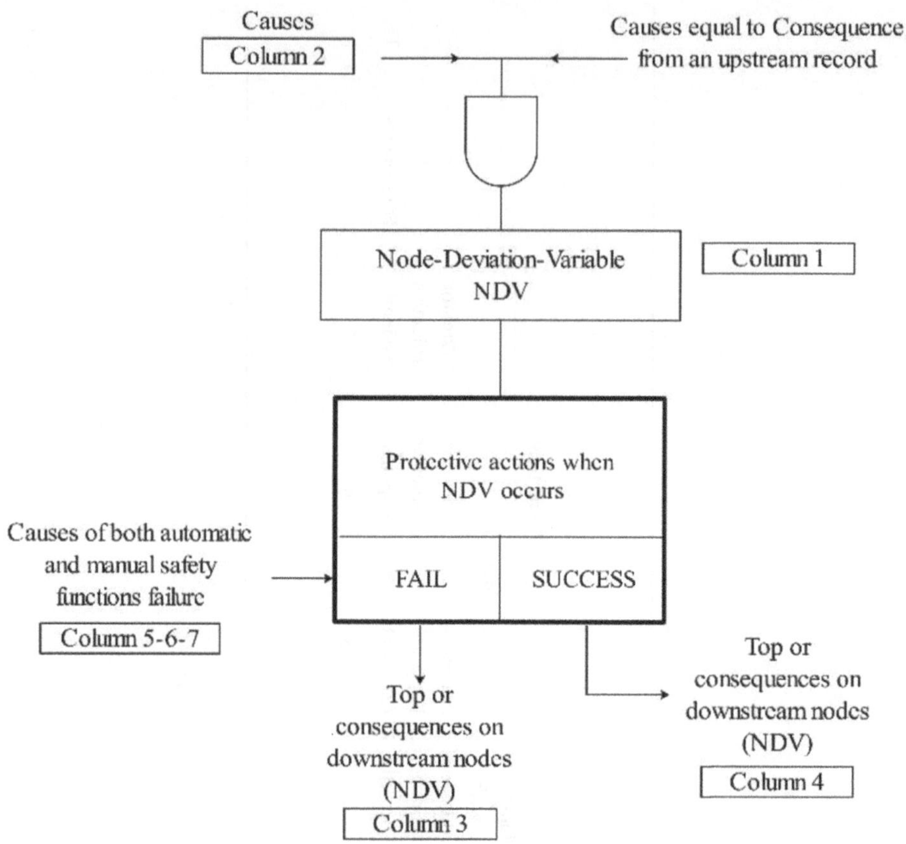

FIGURE 7.3 Cause-Consequence Diagram generation from a ROA-CCD record.

may prove difficult to both interpret and interconnect when the objective is the construction of a comprehensive Fault Tree. For this reason, the use of Incidental Sequence Diagrams, shown in Figure 7.4, represents a more straightforward and practical solution.

The Incidental Sequence Diagram conveys the same set of information as a Cause–Consequence Diagram: however, it is inherently organized in a way that directly supports Fault Tree construction, thereby enabling a more rapid generation of FTs. Within the logical structure of the ISD, the inclusion of Column 4 plays the role of a NOT gate: Column 3 describes the plant condition assuming the failure of protective systems, whereas Column 4 represents the plant condition when protections operate as intended, corresponding to the complementary state. It should be noted that, for the sake of clarity and ease of interpretation, Column 4 may be omitted from the ISD representation when the correct functioning of protective measures does not significantly affect the overall system behavior. Once all Incidental Sequence Diagrams have been generated, they can be systematically interconnected by matching consistent causes and consequences. Through this

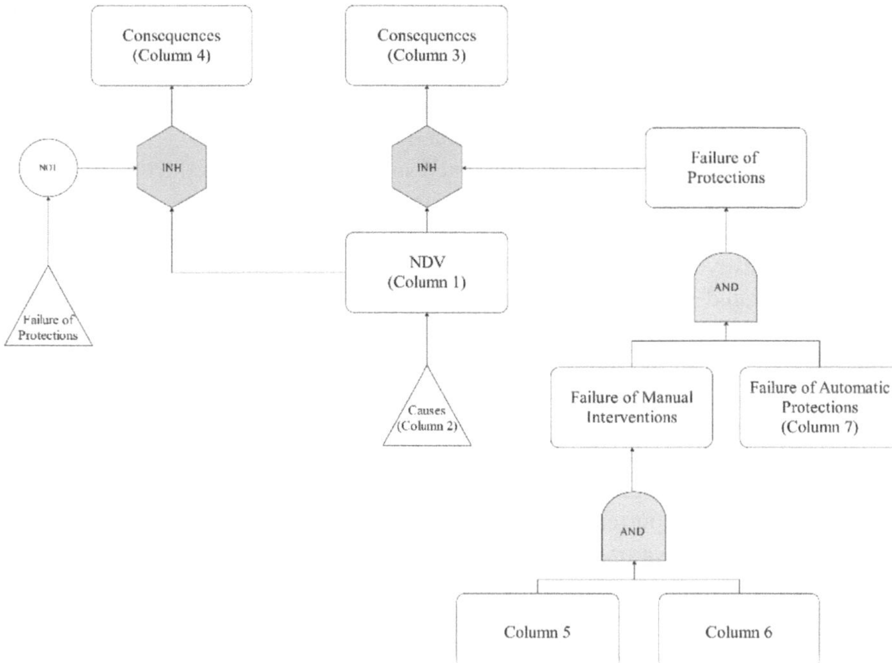

FIGURE 7.4 Incidental Sequences Diagram (ISD) generation from a ROA–CDD record.

process, a complete Fault Tree is obtained for each identified Top Event. In order to solve the Fault Trees, failure or unavailability data for components and probabilities associated with human errors are required, and these parameters can be obtained from the relevant literature sources. The resulting Fault Trees can then be solved and analyzed using specialized software tools.

It should be emphasized that the procedure described in this chapter represents only one of the well-established and widely recognized approaches that can be adopted for Quantitative Risk Analysis. Alternative methodologies, grounded in different theoretical assumptions and levels of complexity, are available and can be applied successfully to perform QRA, depending on the characteristics of the system under investigation, the availability of data, and the objectives of the analysis. The choice of the most appropriate method should, therefore, be guided by the specific context of application rather than by the exclusive adoption of a single analytical approach.

7.2 QUANTITATIVE RISK ASSESSMENT FOR FOOD INDUSTRY APPLICATIONS: THE IMPERIAL SUGAR CASE STUDY

On February 7, 2008, a sugar dust explosion took place on the first floor of the packing building, originating within the enclosed steel belt conveyor located beneath the granulated sugar storage silos [30]. Within seconds, a second, far more extensive dust explosion propagated throughout the entire building. The Imperial Sugar facility was a raw cane sugar refinery, and the accident was thoroughly investigated and reconstructed by the US Chemical Safety and Hazard Investigation Board (CSB). The final CSB report was used as the primary source of information regarding both the process configuration and the explosion dynamics.

At the Imperial Sugar facility, granulated sugar was transported from the refinery to three sugar storage silos through a system composed of screw conveyors, belt conveyors, and bucket elevators. From the silos, the sugar was further conveyed via additional conveyor belts and bucket elevators to specialized sugar processing units and to granulated sugar packaging machines. Sugar products were either packaged within four-story packing buildings surrounding the silos or transferred to railcars and tanker trucks in the bulk sugar loading area [30]. As illustrated in Figure 7.5, the first floor of the packing building housed three reinforced concrete silos, used for conditioning and storing the granulated sugar produced in the refinery. Beneath the silo floor, a conveyor tunnel was installed, containing one Aerobelt® conveyor (responsible for discharging sugar into the west bucket elevator pit) and a steel belt conveyor, which transported granulated sugar from silos 3B and 3C to the packaging production bucket elevator pit located on the eastern side of silo 3C.

In order to minimize potential contamination of the sugar, Imperial Sugar decided in 2007 to enclose the belt conveyor beneath silos 3B and 3C using stainless-steel panels. Less than one year after the installation of this enclosure, a dust explosion occurred within the enclosed conveyor system. Investigations conducted by the CSB revealed that, in addition to inadequate maintenance and housekeeping practices,

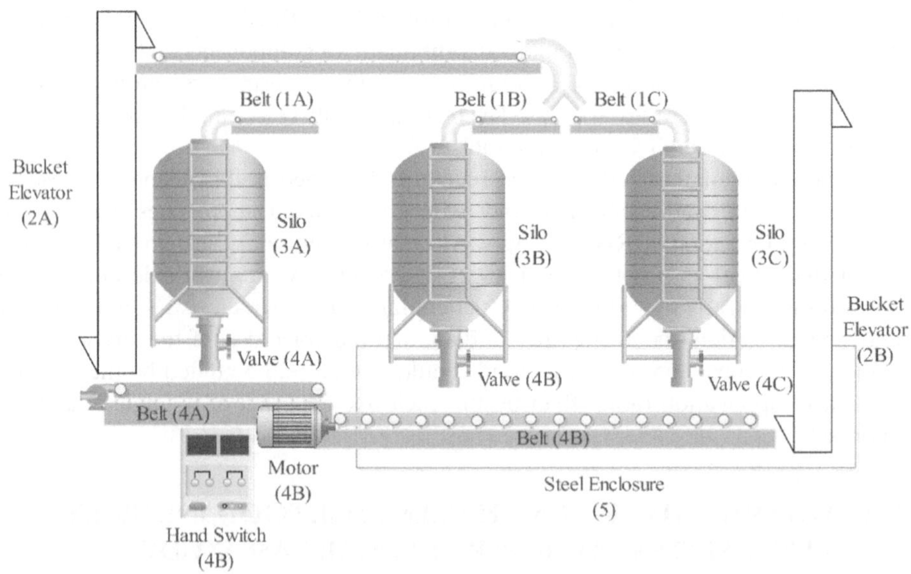

FIGURE 7.5 Granulated sugar supply and discharge through the silos (all related nodes are indicated).

the transport of granulated sugar on the steel belt conveyor led to the formation of airborne sugar dust. Furthermore, interviews with plant personnel indicated that sugar agglomerates frequently obstructed the silo outlet pipes, impeding the flow of sugar onto the conveyor belt. As a result, sugar spilling from the belt was dispersed into the surrounding air in the form of fine dust. The enclosure was not fitted with deflagration venting devices capable of safely relieving overpressure in the event of ignition of airborne sugar dust. In addition, no dust extraction or removal system was installed to prevent the concentration of airborne sugar inside the enclosure from exceeding the Minimum Explosive Concentration, thereby increasing the likelihood of an explosion. CSB investigators also determined that, three days prior to the explosion, workers had been engaged in cleaning operations to remove sugar lumps lodged within silo 3C. During these activities, sugar continued to be conveyed on the steel belt from silo 3B, upstream of the discharge chutes of silo 3C. The accumulated sugar lumps formed a partial obstruction on the conveyor, causing sugar to spill off the belt. Consequently, this release led to the accumulation of spilled sugar within the enclosed conveyor, with portions of the steel belt becoming covered by dust. Within the tunnel, the concentration of airborne sugar dust could exceed the MEC. Based on operator interviews, CSB investigators identified that bearings supporting the steel conveyor rollers occasionally malfunctioned and experienced abnormal heating. From this evidence, it was concluded that a likely ignition source was an overheated bearing within the enclosed conveyor. Elevated temperatures of mechanical components are capable of igniting airborne sugar dust, particularly when the dust remains in prolonged contact with a hot surface, as may occur in a poorly ventilated enclosure. Moreover, if sugar in contact with a hot surface begins to smolder, the combustion gases produced can mix with the airborne dust cloud, effectively lowering the ignition temperature relative to that of pure sugar dust [21].

7.2.1 Application of the ROA Plus-ISD Technique

The first step is the identification of both combustible dust and oxidizing agent: sugar dust is the sole combustible material present in the facility, while oxygen from air acts as the oxidizing agent. The explosive characteristics of sugar dust are summarized in the following: MEC is equal to 115 g/m^3, MIT can range from 360 to 420°C, and MIE is major of 1000 mJ as only granulated sugar was processed. All these parameters may be attained under normal operating conditions within the plant: the minimum explosive concentration can be reached due to resuspension of deposited dust, cleaning operation using an air compressor, or manual handling activities during unclogging operations. The minimum ignition temperature is approximately 400°C; such a temperature can be easily achieved if accumulated sugar deposits start to smolder. On the contrary, the minimum ignition energy is relatively high, but electrical short circuits as well as electric or friction-generated sparks are capable of releasing sufficient energy.

The second step is the identification of all nodes of the plant. As detailed before, it is not mandatory the use of P&IDs, even a simple schematic sketch can be employed successfully. The plant sketch shown in Figure 7.5 corresponds to the section of the plant dedicated to sugar dust collection and conveying operations. The analysis also considers the steel panel installed to protect the lower sugar conveying line. The system is subdivided into five nodes: Node 1A–1B–1C, corresponding to the upper conveyor belts; Node 2A–2B, representing the bucket elevators; Node 3A–3B–3C, comprising the silos; Node 4A–4B–4C, associated with the lower conveyor belts; and Node 5, which includes the steel enclosure. Among the conveying systems, both conventional roller conveyors and Aerobelt® conveyors are employed (1B–1C and 4A). Aerobelt® conveyors operate using air cushions to transport powdered materials and they are specifically designed to reduce leakage and dust dispersion.

The third step is compiling a FMEA table where recovery all failure modes and frequencies of the plant components. The components present in the analyzed section of the plant include electrical switches (Node 4), steel panels (Nodes 1, 3, and 5), conveyor belts (Nodes 1–4), rollers (Nodes 4A, 4B, and 4C), fans associated with the Aerobelt® conveyors (Nodes 4A, 1B, and 1C), and discharge valves (Nodes 3A, 3B, and 3C). In addition, human operators are able to access and interact with Nodes 3, 4, and 5. A simplified FMEA analysis for these components is reported in Table 7.4, with failure rates sourced from the literature when available. Based on the CSB report [30], certain events, such as the generation of an electric spark from a manual switch, can be considered unlikely; however, for simulation purposes, a conservative failure rate of 1e-6 h^{-1} was adopted. Plant data indicate that sugar accumulation may occur at the outlets of silos 3B and 3C, leading to interruption of material flow. In such cases, operators are required to stop the conveyor using a dedicated switch and manually remove the blockage with a steel rod, an operation referred to as "rodding" [30]. Under normal operating conditions, this represents the only planned human intervention in the process.

While it is interesting to analyze the impact of this operation on the process (for example, operators may trigger a friction spark with a misuse of the steel rod), the study should account only for regular protection actions. Therefore, the unique human error considered is related to an unseen sugar clogging due to an "omission of the rodding procedure." The expected failure rate of this omission can be considered as equal to 2E-03 1/y [28].

As already stated, it is necessary to list all the possible ignition sources before going on with the analysis. Based on the plant layout and operating conditions, the most likely ignition sources identified are electrical sparks, friction-generated sparks, hot surfaces, and localized fires (smoldering). Electrical sparks may originate from malfunctioning switches located near the lower conveyor belt line. Friction sparks can be produced by improper execution of the "rodding" procedure by operators or by the impact of falling metallic components. This risk is particularly significant in the bucket elevators, where loosened buckets may detach and strike other metal parts. Hot surfaces or smoldering may result from friction between sugar and roller bearings, or from worn bearings, provided that dust deposits are present. In such cases, ignition may occur if the dust layer reaches its layer ignition temperature (LIT). Unfortunately, no specific tests were conducted to determine the LIT of the sugar handled in the facility. However, according to a Chemwatch Safety Data Sheet for icing sugar, the LIT can be assumed to be approximately 450°C [32].

TABLE 7.4
Simplified FMEA of the Plant Components and Human Errors. Frequencies Are Represented by Failure Rates

Component/Human Error	Failure Mode	Failure Cause	Failure Rate (λ) [31]	$P_i(t)$	ID
4B Hand Switch	Load arc	Electric circuit wearing	Unlikely[30]	8.72E-03	HSEl
	Fails to perform	Switch wearing	4.00E-06 h^{-1}	3.57E-02	HSW
3A-3B-3C silos	Release of sugar	Vessel wearing	3.00E-05 h^{-1}	2.31E-01	3AW-3BW-3CW
	Flow blocked	Sugar clogging	3.00E00 w^{-1}	1.00E+00	3AC-3BC-3CC
3A-3B-3C Discharge valves	Fails to open	Fails to operate	1.00E-03 d^{-1}	3.06E-01	3AVO-3BVO-3CVO
(motor driven)	Fails to close	Rupture	1.00E-04 d^{-1}	3.58E-02	3AVR-3BVR-3CVR
1A-1B-4A	Low flow	Belt wearing	4.00E-05 h^{-1}	2.96E-01	1ABW-1BBW-4ABW
(Aerobelt® conveyor belts)	Fails to start, given run	No air available	5.00E-2 y^{-1}	4.88E-02	1ABNa-1BBWNa-4ABWNa
1A-1B-4A	High air flow	Wrong voltage	1.00E-06 h^{-1}	8.72E-03	1ABWV-1BBWV-4ABWV
(Aerobelt® conveyor fans)		Fan wearing	2.00E-06 h^{-1}	1.74E-02	1ABFW-1BFW-4AFW
	Low air flow	Fan clogged	1.00E-06 h^{1}	8.72E-03	1AFC-1BFC-4AFC
4B	Fails to start, given run	Electric circuit failure	1.00E-06 h^{-1}	8.72E-03	4Bel
(Roller conveyor belt)	Overheating	Roller bearings wearing	5.00E-06 h^{-1}	4.29E-02	4BRW
	Low flow	Belt wearing	4.00E-05 h^{-1}	2.96E-01	4BBW
2A-2B (bucket elevators)	Low flow	Roller wearing	5.00E-06 h^{-1}	4.29E-02	2ARW-2BRW
	Loose bucket	Bucket wearing	1.00E-06 h^{-1}	8.72E-03	2ABW-2ABW
"Rodding"	–	Operator omits procedure	–	2.00E-03	HE
Friction spark	Fall of metal parts	–	Unlikely [30]	8.72E-03	FrS
Enclosure	Enclosure	–	–	1.00-	5En

After the identification of all reasonable ignition sources, all relevant process variables must be summarized in a table (see Table 7.5). Within the plant, sugar flow occurs in silos 3A–3B–3C, from which the material is discharged onto dedicated conveyor belts via the corresponding outlet valves. Sugar transport also takes place in the bucket elevators (Nodes 2A–2B) and along the conveyor belts (Nodes 1A–4A–4B). Sugar mass refers to accumulated deposits, which are mainly expected on conveyor belts 4A–4B, where material may leak through bearing assemblies. Airborne sugar concentration and air temperature are associated with silos 3A–3B–3C and with the enclosure (Node 5). In this context, Nodes 4A–4B are considered closely connected to Node 5, as the conveyors are housed within the enclosure. Pressure is likewise linked to the presence of dust clouds and may occur inside both the silos and the enclosure. The temperature of mechanical components refers to generic equipment parts that may act as hot surfaces capable of igniting dust. Volume is directly related to confinement and it is relevant within the silos and the enclosure.

After all these preliminary steps, the ROA Plus–ISD methodology was applied to the facility under investigation. In the following, ROA table related to both the steel conveyor belt (Node 4B), and the enclosure (Node 5) (see Table 7.6), which represent the most critical areas of the plant as they correspond to the zone where the steel enclosure was installed and where the primary explosion occurred [30], will be reported. To improve readability, deviations deemed insignificant or redundant were excluded from the table. For instance, the deviation associated with a low temperature of mechanical components is considered highly unlikely, and it leads only to negligible consequences; therefore, deviations involving this condition were not included.

Based on the ROA table, three main Top Events (TEs) can be identified: sugar dust cloud formation (TE1), sugar smoldering (TE2), and sugar dust explosion (TE3). The main deviations associated with these three TEs are elevated dust concentration, increased air temperature and pressure, confinement, and the presence of ignition sources (high metallic part temperatures and sparks). Under normal operating conditions, sugar dust may become airborne as a result of routine process activities, including material transport along conveyor belts and sugar discharge from silo

TABLE 7.5
Process Variables Analyzed in the ROA Plus–ISD Application

Symbol	Variable Deviation
F	Sugar Flow
m	Sugar mass
C_a	Airborne sugar concentration
T	Temperature of mechanical parts
T_a	Air Temperature
P	Pressure (within the enclosure)
V	Volume (available, ref. enclosure)

TABLE 7.6
ROA for the Enclosed Steel Belt Conveyor (Node 4B) and the Enclosure (Node 5)

					Protections				
					Manual				
REC	NDV	Causes	Consequences due to Protection Failure	Plant State with Protections Working Correctly	Alarm (Optical/ Acoustic)	Operator actions on Components	Automatic Protective Systems	Notes	TE
4.1	4B.h.F	Discharge Valve 3B Rupture OR Discharge Valve 3C Rupture	4B.h.m (locally on or near the belt)	4B.h.m (locally on or near the belt)	–	–	–	Sugar spillage.	
4.2	4B.l.F	Sugar Clogging 3B OR Sugar Clogging 3C OR 4B belt fails to start (electrical circuit failure)	4B.h.m (locally on or near the belt)	System goes back to normal funztioning	–	Belt switch off AND Rodding operation	–	Sugar blocking locally.	

(Continued)

TABLE 7.6 CONTINUED
ROA for the Enclosed Steel Belt Conveyor (Node 4B) and the Enclosure (Node 5)

					Protections				
					Manual				
REC	NDV	Causes	Consequences due to Protection Failure	Plant State with Protections Working Correctly	Alarm (Optical/ Acoustic)	Operator actions on Components	Automatic Protective Systems	Notes	TE
4.3	4B.h.m (locally on or near the belt)	Rec 4.1 OR Rec 4.2 OR Wearing of 4B Belt OR 3B Vessel wearing OR 3C Vessel wearing	5.h.Ca (> MEC)	5.h.Ca (>MEC)	–	–	–	Sugar may slide onto the bearings and disperse in air due to the falling out of the belt.	
4.4	5.h.Ca (>MEC)	Rec 4.3 AND 5.l.V (confinement)	Explosive Dust cloud formation.	Explosive Dust cloud formation.	–	–	–		TE1
4.5	4B.h.T (higher than normal but lower than LIT)	Roller wearing	4B.hh.T (higher than LIT)	4B.hh.T (higher than LIT)	–	–	–	–	

4.6	4B.hh.T (higher than LIT)	(Rec 4.3 AND Rec 4.5) OR (Rec 4.3 AND Friction Spark) OR (Rec 4.3 AND Electric Spark)	Smoldering (Local Fire) $5.h.T_a$ (higher than MIT)	Smoldering (Local Fire) $5.h.T_a$ (higher than MIT)	–	–	–	TE2
4.7	5.h.Ta (higher than MIT)	Rec 4.6	5.h.P	5.h.P	–	–	–	
4.8	5.h.P	(Rec 4.7 OR Electric Spark OR Friction Spark) AND Rec 4.4	Enclosure explosion	Enclosure explosion	–	–	–	TE3

outlets. Consequently, high dust concentrations are directly linked to the presence of accumulated sugar deposits onto the belt.

Additional mechanisms capable of dispersing sugar dust into the air include human activities, such as cleaning with compressed air or the "rodding" procedure, during which operators remove blockages at silo outlets following clogging events. While compressed-air cleaning is not carried out in this section of the plant (because of safety reasons), the rodding operation is considered as a protective measure rather than a mechanism through which it is possible to create a hazard (this is reflected in the ROA table).

With regard to process variables, a high airborne sugar concentration (h.C_a) corresponds to levels exceeding the minimum explosive concentration (MEC). Elevated mechanical component temperatures (h.T) indicate temperatures below the layer ignition temperature (LIT) but above normal operating conditions, whereas very high temperatures (hh.T) exceed the LIT and may lead to ignition of sugar deposits. Reduced volume (l.V) denotes the presence of confinement, which promotes both dust accumulation and pressure buildup; in this case, confinement is provided by the enclosure. In the absence of the enclosure, high dust concentrations would be limited to localized areas, such as inside the silos or near sugar deposits and discharge points.

Although explosions within silos represent a significant hazard (addressed separately in the analysis of individual silo nodes), for Nodes 4B and 5, the removal of the enclosure would primarily limit the risk to localized fires.

As previously discussed, the most significant ignition sources were identified as elevated temperatures resulting from a sugar fire (or smoldering), defined as temperatures exceeding the minimum ignition temperature (h.T_a), as well as electrical and friction-generated sparks. High-temperature scenarios are adequately captured by standard deviations, whereas electrical and friction sparks do not necessarily constitute basic events (even if, in this example, they are of this type for the sake of simplicity) and, therefore, should require separate development. With respect to the equipment, electrical sparks may originate from a malfunctioning manual switch, as highlighted in the FMEA analysis (Table 7.4); however, this event has been classified as unlikely [30]. Regarding the overpressure effects, literature data indicate that sugar dust exhibits deflagration indices ranging from 30 to 150 bar·m/s [30, 33], depending on moisture content and particle size distribution. These values suggest that sugar dust explosions can cause severe damage to both equipment and personnel. Consequently, a pressure level of concern equal to 2.76 kPa, corresponding to the potential collapse of the enclosure, can be adopted [31].

7.2.2 Fault Trees Analysis and Risk Evaluation

Based on the ROA table (see Table 7.6), ISDs can be generated automatically and merged into the corresponding TEs fault trees.

Figures from 7.6 to 7.8 report the complete fault trees, generated by linking all the related ISDs. Figure 7.9 reports the event 4B.h.m separately using a transfer gate. All these FTs can now be analyzed with a dedicated software, such as OpenFTA. This software performs fault tree analysis and provides outputs such as minimal cut sets,

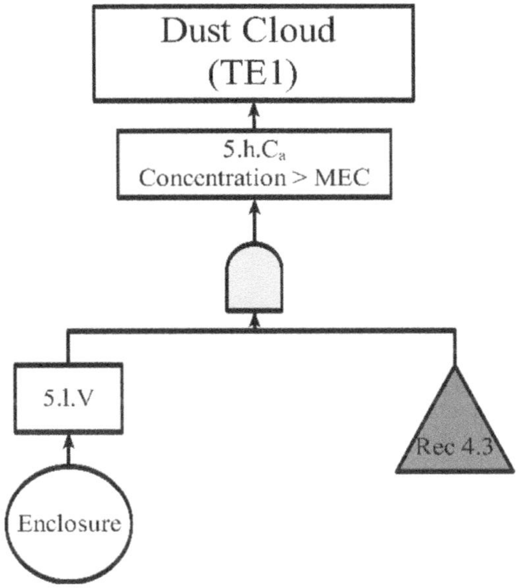

FIGURE 7.6 Fault tree for TE1: sugar dust cloud formation.

event probabilities, and importance measures for individual basic events. To obtain reliable probability estimates, both the mission time and component unreliability functions must be defined. A mission time of one year was assumed, which is considered reasonable given the recent introduction of a new element in the process, namely the steel enclosure. Under this assumption, maintenance and repair activities were regarded as negligible. Accordingly, the unavailability of the i-th component was estimated using a Poisson distribution [34].

For human errors, probability values were directly obtained from available databases. The probabilities associated with all components listed in the FMEA are reported in Table 7.4. Events classified as unlikely were modeled with a failure rate of 1E-06 1/h, resulting in an annual probability of 8.72E-03. The presence of the enclosure was treated as a Boolean variable, and once implemented, its associated probability was set equal to unity.

For the dust cloud top event, eight primary events were identified, as illustrated in Figure 7.6, resulting in seven minimal cut sets (MCSs). Among these, three are second-order MCSs, while the remaining four are of third order. The local fire top event (TE2) involves 13 primary events, with a total of 30 minimal cut sets identified; 18 of these are second order, whereas the remaining 12 are third-order MCSs. In the case of the dust explosion analysis (TE3), 14 primary events were considered, leading to the identification of 21 minimal cut sets. Of these, nine are third order and 12 are fourth order MCSs. A summary of all results is provided in Table 7.7.

From a general perspective, a low number of minimal cut sets combined with higher MCS order is typically indicative of an inherently safer process. In the present

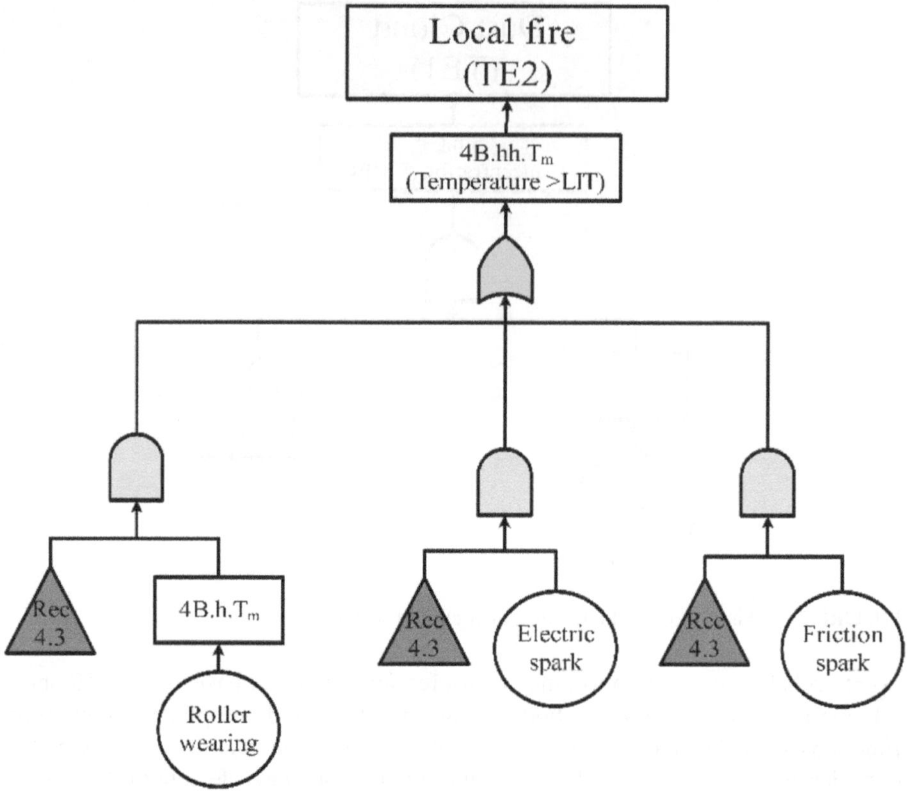

FIGURE 7.7 Fault tree for TE2: smoldering (local fire).

analysis, a relatively large number of MCSs is associated with the local fire and dust explosion top events. This outcome is largely attributable to the ROA Plus–ISDs methodology, which requires the consideration of all potential ignition sources. The relatively low MCS orders (ranging from two to four) suggest limited protective measures; however, the most severe scenario, namely the dust explosion, exhibits the highest MCS orders. The estimated probabilities of the top events are 1.13E−01 for dust cloud formation, 2.75E−02 for local fire, and 6.73E−03 for dust explosion. These values correspond to expected occurrence intervals of approximately 8.8 years for a dust cloud, 36 years for a local fire, and 148 years for a dust explosion.

The obtained results are internally consistent, as an order-of-magnitude difference between dust cloud formation and local fire, as well as between local fire and dust explosion, is reasonably expected. According to these findings, the formation of a dust cloud appears to be a relatively frequent event, a conclusion supported by the accident reconstruction, which reported sugar dispersion throughout the facility [30]. The dominant contributors to dust cloud formation are the presence of the enclosure, as it appears in all minimal cut sets, and failure of the manual switch during sugar clogging events (with an importance of 63.19%). Conversely, operator omission of

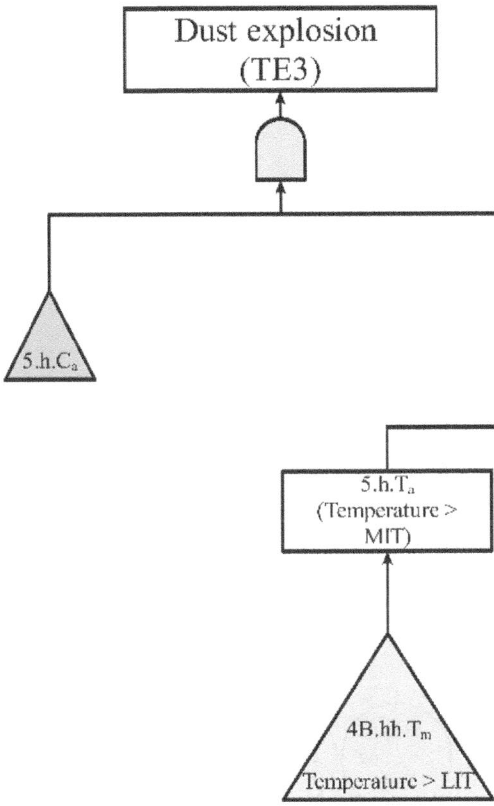

FIGURE 7.8 Fault tree for TE3: dust explosion in the enclosure.

procedures (3.53% importance) and failure of the 4B conveyor belt to start (7.70% importance) have a comparatively minor influence. In the case of local fire scenarios, the most significant contributing factor is wearing of the 4B belt bearings, with an importance of 94.27%. This observation is particularly noteworthy, as it is strongly supported by accident investigations in which operators reported frequent overheating of bearings during operation [30]. The least influential primary events remain operator procedural omissions (0.88% importance) and failure of the 4B belt to start (1.91% importance). For dust explosion scenarios, the enclosure again exhibits 100% importance, as it is the sole source of confinement at this node, while bearing wear continues to represent the most likely ignition source, with an importance of 99.24%. Such levels of risk are considered unacceptable under the ALARP principle [34]. The results were also compared with a separate risk assessment conducted on the same facility by Abuswer et al. [24]. In that study, the probability of a dust explosion was adjusted using a correction factor accounting for explosion consequences, leading to a reduction from 3.76E−02 to 4.40E−05 [24]. The probability estimated in the present work lies between these values, at 6.73E−03, and closely matches the result

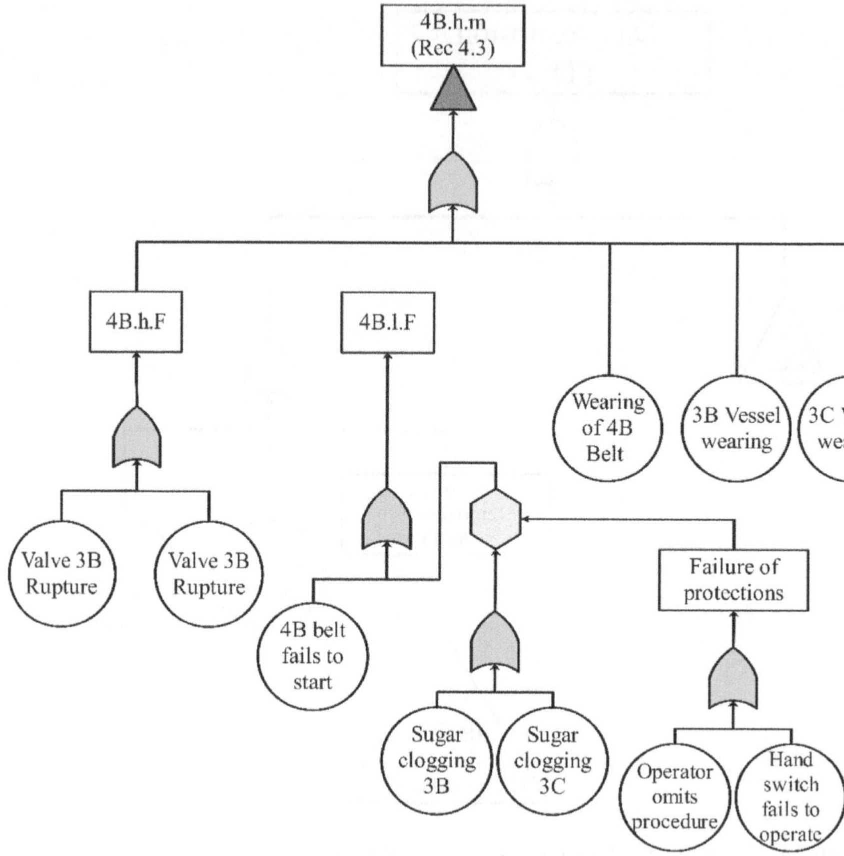

FIGURE 7.9 Transfer gate for the 4B.h.m deviation (Rec 4.3).

obtained when no QRMF correction is applied. Although the present analysis does not address event severity, it is important to note that fires and explosions generate individual risk indices that cannot be fully characterized by probability values alone. In this sense, local fires, and dust explosions with their probabilities of occurrence will not generate individual risk levels lower than 1.0E-03, which is the lower bound for the ALARP region. This analysis highlights the importance of considering the effect of plant changes, such as the addition of an enclosure on conveyor belts, by performing a proper risk assessment.

7.3 QUANTITATIVE RISK ASSESSMENT FOR PHARMA INDUSTRY APPLICATIONS: THE CASE OF A FLUID BED GRANULATOR

Granulation of active pharmaceutical ingredients is one of the most important steps in the preparation of a final product. Such an operation is mainly done in fluidized bed granulators and implies a series of safety concerns as detailed in Chapter 6. In

TABLE 7.7
Probabilities of Occurrence P for Every MCS Identified (refer to Table 7.4 for IDs)

Dust Cloud (TE1)			Local Fire (TE2)			Dust Explosion (TE3)		
#	MCS	P[-]	#	MCS	P[-]	#	MCS	P[-]
1	3BVR,5En	3.58E-02	1	3BVR,4BRW	1.53E-03	1	3BVR,4BRW,5En	1.53E-03
2	3CVR,5En	3.58E-02	2	3BVR,HSEl	3.12E-04	2	3BVR,HSEl,5En	3.12E-04
3	4BEl,5En	8.71E-03	3	3BVR,FrS	3.12E-04	3	3BVR,FrS,5En	3.12E-04
4	3BC,5En,HSW	3.57E-02	4	3BW,4BRW	9.01E-03	4	3CVR,4BRW,5En	1.53E-03
5	3BC,5En,HE	2.00E-03	5	3BW,HSEl	1.83E-03	5	3CVR,HSEl,5En	3.12E-04
6	3CC,5En,HSW	3.57E-02	6	3BW,FrS	1.83E-03	6	3CVR,FrS,5En	3.12E-04
7	3CC,5En,HE	2.00E-03	7	3CVR,4BRW	1.53E-03	7	4BEl,4BRW,5En	3.74E-04
			8	3CVR,HSEl	3.12E-04	8	4BEl,HSEl,5En	7.60E-05
			9	3CVR,FrS	3.12E-04	9	4BEl,FrS,5En	7.60E-05
			10	3CW,4BRW	9.01E-03	10	3BC,4BRW,5En,HSW	1.53E-03
			11	3CW,HSEl	1.83E-03	11	3BC,4BRW,5En,HE	8.56E-05
			12	3CW,FrS	1.83E-03	12	3BC,HSEl,5En,HSW	3.11E-04
			13	4BBW,4BRW	1.27E-03	13	3BC,HSEl,5En,HE	1.74E-05
			14	4BRW,4BEl	3.74E-04	14	3BC,FrS,5En,HSW	3.11E-04
			15	4BBW,HSEl	2.58E-04	15	3BC,FrS,5En,HE	1.74E-05
			16	4BBW,FrS	2.58E-04	16	3CC,4BRW,5En,HSW	1.53E-03
			17	4BEl,HSEl	7.60E-05	17	3CC,4BRW,5En,HE	8.56E-05
			18	4BEl,FrS	7.60E-05	18	3CC,HSEl,5En,HSW	3.11E-04
			19	3BC,4BRW,HSW	1.53E-03	19	3CC,HSEl,5En,HE	1.74E-05
			20	3BC,4BRW,HE	8.57E-05	20	3CC,FrS,5En,HSW	3.11E-04
			21	3BC,HSEl,HSW	3.12E-04	21	3CC,FrS,5En,HE	1.74E-05

(*Continued*)

TABLE 7.7 CONTINUED
Probabilities of Occurrence P for Every MCS Identified (refer to Table 7.4 for IDs)

Dust Cloud (TE1)			Local Fire (TE2)			Dust Explosion (TE3)		
#	MCS	P[-]	#	MCS	P[-]	#	MCS	P[-]
			22	3BC,HSEl,HE	1.74E-05			
			23	3BC,FrS,HSW	3.12E-04			
			24	3BC,FrS,HE	1.74E-05			
			25	3CC,4BRW,HSW	1.53E-03			
			26	3CC,4BRW,HE	8.57E-05			
			27	3CC,HSEl,HSW	3.12E-04			
			28	3CC,HSEl,HE	1.74E-05			
			29	3CC,FrS,HSW	3.12E-04			
			30	3CC,FrS,HE	1.74E-05			

the following, a generic unit performing such an operation with a potentially explosive powder will be analyzed through the application of both a Recursive Operability Analysis and a Fault Tree Analysis followed by a risk evaluation.

7.3.1 Description of the Plant Components

A generic Fluidized Bed Granulator (FBG) unit can be divided into the following major subsystems:

- **Air preparation unit.** Process air is typically conditioned through a sequence of treatment steps, including pre-filtration, pre-heating, dehumidifying, re-humidification, and high-efficiency particulate filtration. The system uses fresh outdoor air which, for pharmaceutical operations, must be free of dust and other contaminants. Experience shows that process-air quality can strongly affect granulation performance; therefore, accurate and continuous monitoring of air properties is essential.
- **Powder container and air distributor.** Once air has been brought to the required operating conditions, it must be uniformly distributed through the powder bed. This step is critical: non-uniform air distribution may cause flow channeling and unstable or incomplete fluidization. For reliable operation, both the product container and the air distribution system must be properly designed and optimized for the specific formulation being processed.
- **Spray nozzles.** The nozzles atomize the binder (in this case, water), which is then injected into the fluidized bed. The goal is to generate a spray, i.e., a dispersion of fine liquid droplets in a gas stream. Because droplets have a high specific surface area, they interact efficiently with the powder, enabling controlled agglomeration. Spray generation can be achieved through different mechanisms. In this example, the liquid is drawn into the nozzle by a pressure drop, and a compressed-air jet breaks the liquid stream into droplets.
- **Expansion (calming) chamber and filters.** The exhaust air leaving the process chamber must be separated from residual dust. Coarser particles fall back into the bed by gravity, while finer particles are retained by a bag filter system (filter sleeves).
- **Exhaust fan.** At the end of the process, air is discharged through an exhaust fan, whose main purpose is to maintain a slight negative pressure inside the system, ensuring continuous inward leakage of atmospheric air rather than outward release of dust. Air flow is controlled by valves installed upstream of the fan.
- **Control system.** The control architecture consists of a PLC (Programmable Logic Controller) for machine control functions and a PC running software which provides the operator interface. The PLC and PC exchange information continuously throughout the process: the system downloads nominal values (setpoints) and receives real-time process measurements. The operator display in the production area shows the machine flow

diagrams and the progress status, allowing detailed monitoring of each component; for example, the current position of inlet/outlet duct valves and modulation flaps.

The complete granulation plant is installed across two floors, while only the terminal section of the ductwork, including the overpressure vent panel, is located on the roof. Before describing the plant components in detail, a synthetic P&ID of the plant is provided (Figure 7.10).

The following list (Table 7.8) summarizes the plant components, grouped by process variable (or function) and their location.

Only the components most relevant to the applied Recursive Operability Analysis are described in detail in the following.

(1) **Inlet air system: W102 main heater (heating group).** The purpose of the heating group is to maintain a stable inlet air temperature according to the setpoint. As shown in the plant scheme, W102 is a heat exchanger equipped with a condensate collection system and a solenoid valve (V101) regulating steam flow.
(2) **TIC101: inlet air temperature control/indication.** TIC101 is the temperature controller and indicator associated with the heating group. It measures

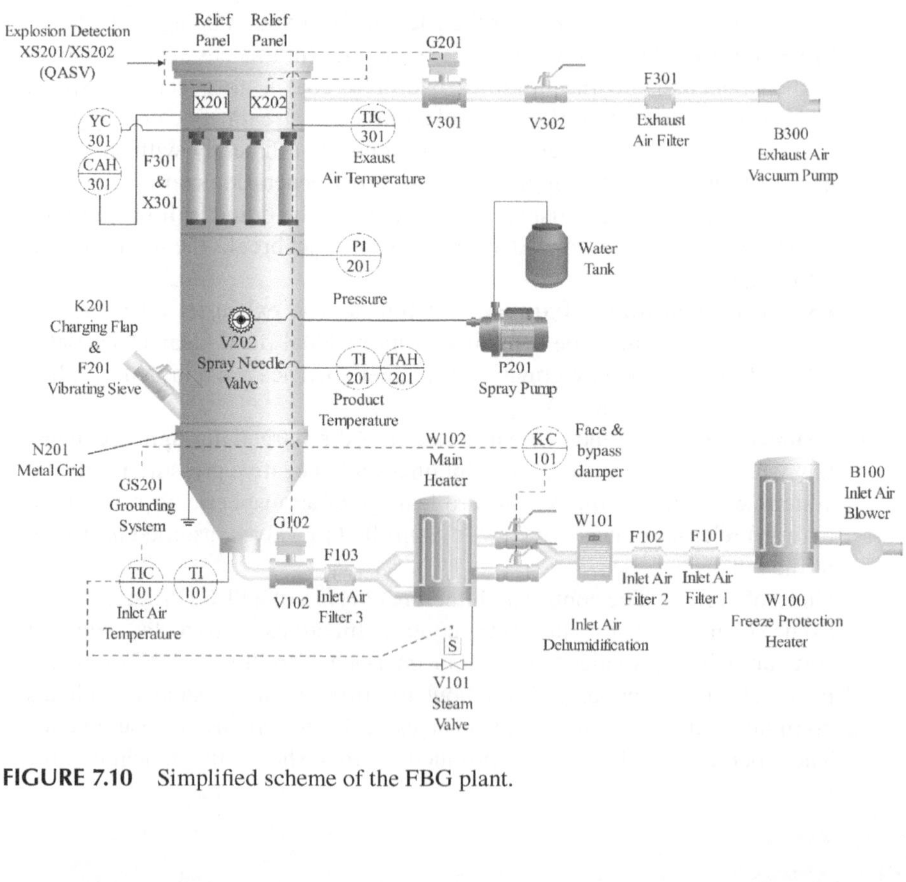

FIGURE 7.10 Simplified scheme of the FBG plant.

TABLE 7.8
List of the Components of the Plant

Component	Variable or Typology	Location
W100: Freeze Protection Heater	Temperature	Inlet Air System
B100: Inlet Air Blower	Flux	Inlet Air System
F101: Inlet Air Filter 1	Filter	Inlet Air System
F102: Inlet Air Filter 2		
F103: Inlet Air Filter 3		
W101: Inlet Air Dehumidification	De-humidification	Inlet Air System
W102: Main heater	Temperature	Inlet Air System
V102: Inlet Air QASV	Protection	Inlet Air System
Explosion Detection (QASV) IN/OUT	Protection	Inlet/Exhaust Air System
KC101: Face and bypass damper	Flux	Inlet Air System
TI101/TIC101: Inlet Air Temperature Indicator/Controller	Temperature	Inlet Air System
V101: Steam Valve	Temperature	Inlet Air System
T201: Product Temperature	Temperature	Machine Tower
K201: Charging Flap	API Loading	Machine Tower
F201: Vibrating Sieve	API Loading & Screening	Machine Tower
X301: Product Filter Sealing	Bag Filter Sealing	Machine Tower
F301: Product Filter/Long Stroke Cylinder	Bag Filter Support	Machine Tower
YC301: Shaking Cylinder	Bag Filter Cleaning	Machine Tower
PI201: Differential Pressure	Pressure	Machine Tower
GS201: Grounding System	Electrostatic Discharges	Machine Tower
T301: Exhaust Air Temperature	Temperature	Exhaust Air System
V301: Exhaust Air QASV	Protection	Exhaust Air System
V302: Exhaust Air Modulating flap	Flux	Exhaust Air System
F301: Exhaust Air Filter	Filter	Exhaust Air System
X201/ X202: Explosion Relief Panel	Safety Device	Exhaust Air System
XS201/ XS202: Explosion Relief Flap	Safety Sensors	Exhaust Air System
P201: Spray Pump	Pump	Spray System
V202: Spray Needle Valve	Flux	Spray System
G201/G102: Exhaust/Inlet quick-acting shut-off valves switches	Protection	Exhaust/Inlet Air System
N201: Metal Grid (Product Containment)	Support	Machine Tower
B300: Exhaust Air Vacuum Pump	Flux	Exhaust Air System
CAH301: High Dust Concentration Alarm	Protection	Exhaust Air System
TAH201: High Product Temperature Alarm	Protection	Machine Tower
Operator Interface Terminal (OIT)	Software	Control

Default positions/states: (a) Flaps: closed; (b) Valves: closed; and (c) Motor (fan): off.

inlet air temperature and transmits the signal to the PLC, enabling real-time display on the operator screen. In automatic mode, the measured temperature and the inlet temperature setpoint are used to regulate the position of the face and bypass damper K101 via a PID control loop.

(3) **KC101: face and bypass damper.** The face and bypass damper separates the incoming air stream into two branches, one passing through the heater and one bypassing it. Based on the output of the TIC101 controller, the two flaps move synchronously to ensure effective mixing of heated and unheated air.

(4) **V101: steam valve.** The pneumatic signal generated by the TIC101 controller is converted into an electrical ON/OFF command, which drives the solenoid steam valve V101, opening or closing the steam supply to the heating circuit.

(5) **TI201: product temperature.** The TI201 sensor measures product temperature inside the FBG unit and sends the value to the PLC for display and monitoring.

(6) **Explosion detection and isolation: inlet/outlet.** To prevent equipment damage due to a potential dust explosion and, in particular, to prevent flame propagation along the inlet and exhaust ducts, the system is equipped with two quick-acting shut-off valves (QASV): V201 and V102. In the event of an explosion, these valves close almost instantaneously, driven by the high pressure in the associated nitrogen tank, which actuates the valve mechanism. Their default position is open. If the explosion relief panels X201 and X202, located above the filter sleeves, open due to overpressure, switches G201 and G102 are activated. This signal triggers the release of high-pressure nitrogen to the QASV actuator, closing the duct.

In the following, a brief overview of the plant functioning is provided to better understand all the further steps of the QRA.

Concerning the process, it is well established that during the production, transport, or handling of powders, electrostatic charges may be generated due to rapid particle movement through equipment surfaces. Charges may accumulate in the powder bulk, in airborne dust clouds, or in deposited layers, and may also be induced on equipment surfaces if these are not properly grounded. Ignition risk arises during charge recombination, as electrostatic discharge can release enough energy to ignite a combustible dust cloud. For deposited powders, ignition risk is generally lower, but stored charge may persist depending on dielectric constant and conductivity. In this plant, the most relevant electrostatic hazard would occur when charged particles are suspended, such as during pneumatic conveying. For powder handling, two different systems are used: a vibrating sieve and a bin blender, both operating under vacuum. The vibrating sieve is used for screening and feeding API into the FBG (and, separately, for transferring flavors to the bin), while the bin blender is used during product discharge from the fluid bed. The vibrating sieve conveys powder through flexible plastic hoses fitted with a copper wire spiral along their entire length to improve conductivity. All equipment is grounded to dissipate any accumulated charge. On the roof above the FBG unit, there is no conventional lightning rod; instead, a lightning protection mesh is installed. Inspection of the mesh and grounding is required every ten years. Inside the building, several surge protection devices (Type I, II, III) are installed with different attenuation capabilities. Unexpected plant shutdowns due to control failures or loss of industrial data transmission can be extremely costly, not only due to repair costs but also because of downtime. Such events reduce the mean time between failures (MTBF). Overvoltage hazards are significant, as demonstrated by insurance loss statistics; in general, all electrical production equipment, from control cabinets to small electronic components, is threatened by overvoltage events. For this reason, surge protection has become increasingly important as components become smaller and automation becomes more widespread. Surge arresters with extremely fast response times are used to limit voltage peaks before hazardous conditions develop. Their response time is on the order of nanoseconds, and they must withstand very high currents (potentially several thousand amperes) while maintaining low residual voltage. Effective surge protection is characterized by fast response, high current capacity, low residual voltage, and rapid electrical recovery after grounding.

The dust alarm (CAH301) triggered by the opacity meter above the filter sleeves can frequently occur during the early phase of the process. At this stage, higher pressures can resuspend dust deposited in the exhaust ductwork near the opacity meter. When this alarm is triggered, the opacity meter is cleaned immediately. The alarm may also occur if filter sleeves are damaged, which can happen during the shaking cycles. In practice, the alarm is expected when the campaign starts or is near the end of discharge, when the finest powders remain in the unit and may pass through the filters. Filter sleeves should be replaced once per week, or twice if required. If a dust-related alarm occurs, it must be reported when completing the batch checklist.

If alarms occur during heating or cooling, supervisors are notified and the technical office is contacted for checks; however, after acknowledging the alarm, the process can continue without major issues. If the problem occurs during the wetting phase, the production manager must be informed, and additional laboratory tests are required. Quality Assurance (QA) then decides whether the product remains acceptable for packaging. Because all actions are recorded (date, time, operator), QA reviews any alarms triggered during the batch before making a decision.

Operators may be exposed to multiple acoustic signals from different sources (bin washer, drying oven, warehouse and department doors, fire alarms, nearby machines, and the FBG). The FBG control system also includes a feature that can send alarms directly to the operator's mobile phone, which is useful if the operator temporarily leaves the area.

When an alarm occurs, an icon appears on the OIT screen, and an audible signal indicates an impending malfunction. The icon remains active until acknowledged by the operator. Alarms may have a main alarm and sub-alarms; the highest-priority alarm is always displayed first. After acknowledgement, any sub-alarms that may have triggered the main alarm are shown. The control system monitors critical parameters and generates an alarm only when values exceed the defined upper or lower tolerance limits.

Two time-based evaluations are used to determine whether an alarm should be generated:

- Alarm Settling Time (ts): the initial time window at the start of a new operating phase during which alarms are suppressed because parameters may still be stabilizing and not yet within tolerance.
- Alarm Delay Time (td): the maximum time allowed for parameters to return within tolerance; if values remain out of range beyond this window, the alarm is triggered.

In hazardous situations, the machine can be stopped using the emergency stop button.

The plant includes protection systems intended primarily to mitigate the consequences of a dust explosion: explosion containment, plant segregation/isolation, explosion suppression, and explosion venting. In the FBG unit, the adopted safety devices include: two fast-acting explosion isolation valves installed on the inlet and exhaust lines, two venting panels (above the filter sleeves) designed to open in the direction of venting to atmosphere, one additional vent panel installed on the roof (not represented in the scheme for the sake of simplicity).

Inspection frequencies and maintenance tasks for both safety-critical and process components of the FBG unit are defined in the technical documentation provided with the equipment. Some tasks can be performed by the company's maintenance team, whereas others must be carried out by the manufacturer or by personnel certified by the manufacturer. According to the ATEX directive, the user's personnel cannot be authorized by the manufacturer to perform certain operations on safety devices, since the required qualifications are very stringent and potential conflicts of interest must be avoided.

It is necessary to ensure good condition not only of primary safety devices (e.g., QASV valves and explosion relief panels), but also of other parts such as inflatable seals, hydraulic clamping systems, sight glasses, flexible transfer hoses, and many others (not explicitly referred in this example). Inspection and maintenance frequency depends mainly on product properties, production plans, and GMP requirements defined by the user and implemented within the maintenance program.

Two types of routine maintenance are distinguished: preventive and predictive. Preventive maintenance is valuable, but it exhibits limitations: under normal operation, variable conditions such as temperature and vibration may accelerate or slow down equipment degradation, which fixed preventive schedules may not adequately capture. As a result, equipment may be stopped or replaced prematurely, causing unnecessary maintenance costs and production losses. Conversely, if equipment operates under unusually stressful conditions, it may fail earlier than expected; in such cases, predictive maintenance is applied.

7.3.2 Description of the Granulation Process

Fluidized-bed granulation (FBG) is a process in which granules are formed within a single piece of equipment, where a binder (generally water) is atomized onto a bed of fluidized powder. The first industrial applications of this technique date back to 1942, and they were developed by the Standard Oil Company (now known as ExxonMobil). A generic granulation process of an API can be divided into the following stages:

- **Pre-heating.** The equipment is preheated using a given air flow rate at a temperature of 45–50°C.
- **Periodic filter shaking.** The upper bag filters located in the top section of the granulator are periodically shaken throughout the entire process. Each shaking cycle lasts some minutes (generally, 1–2 min).
- **Loading.** The API is introduced into the granulator through a charging lance and a vibrating sieve equipped with a given mesh size. During the loading phase, the operator intermittently stops and restarts the air suction. This is necessary because the FBG charging butterfly valve tends to partially close during filter shaking; temporarily stopping the sieve prevents powder compaction on the sieve screen. Once powder loading is complete, the charging valve is closed. Any residual API granules remaining on the sieve mesh are manually crushed, re-sieved, and finally removed using a dedicated vacuum system.
- **Heating.** The API is heated until a product temperature of 45–48°C is reached. This is achieved by supplying air at 52–54°C with a given flow rate. During this phase, bag filter shaking cycles of 15–20 seconds occur every two minutes.
- **Wetting (spraying).** This is the longest and most critical stage of the entire process. A vessel containing demineralized water is connected to the FBG unit, with the spray nozzle positioned at the highest available spraying point on the granulator wall. The water is sprayed into the chamber for a given period at a pre-determined flow rate. A photoelectric sensor detects both the absence of water in the outlet tubing of the peristaltic pump and the presence of air in the liquid line. The operator monitors the water level via an Operator Interface Terminal (OIT) and performs additional visual checks directly at the vessel. At the same time, a controlled air flow is supplied to ensure continuous mixing of the powder bed. Filter shaking continues at two-minute intervals.
- **Drying.** The process is maintained under constant fluidization for approximately 20 minutes, increasing both the inlet air flow rate and temperature (70–72°C). Drying proceeds until the granules reach a product temperature of 60–62°C.

- **Cooling.** The inlet air temperature is then reduced to ambient conditions to cool the granulate. This phase continues until the product temperature falls below 40–42°C.
- **Discharging.** Finally, the granulated product is discharged by connecting a hose from the FBG discharge valve to a unit which pneumatically conveys the product into the designated bin.

Throughout the process, the operator can monitor the system status via the OIT, which displays the current process phase and all relevant parameters. Continuous supervision is not strictly required, as the system is equipped with sensors that automatically trigger alarms whenever abnormal operating conditions are detected.

7.3.3 FMEA-Like Database of Failure Rates and Flammability Explosibility Data Collection

The safety team's work should begin with a detailed review of the FBG process, including an assessment of the P&ID (if available) and the associated process-variable control systems. Three key variables are identified as the focus of the study: dust concentration, temperature, and pressure. The instrumentation and control scheme can be divided into three main sections:

- inlet air line;
- exhaust air line;
- water/binder line (from the vessel to the spray nozzle).

Next, a FMEA-like table must be compiled. It is termed "FMEA-like," because it includes the same core fields as a standard FMEA, but it excludes the columns related to consequences, recommended actions, and criticality. This structure can be chosen to facilitate the development of the ROA method. Table 7.9 presents the FMEA-like worksheet and lists the failure causes of the instruments relevant to the ROA application (i.e., the critical components whose malfunction can lead to system failure): temperature controller, temperature transmitter, flapper (damper), and solenoid valve. In the column "Failure frequency (per failure mode) [1/h]," the data sources (databases) used to derive each value are reported in parentheses.

Let's consider a generic API powder with the following characteristics:

- Particle size distribution (d50): 100–150 μm.
- K_{st}: 88 bar/(m s) @ 1250 g/m^3.
- MIE: >500.
- LOC: 11%.
- Chemical composition: epoxy group present. Melting with exothermic reactivity can be traced using Differential Scanning Calorimetry (DSC) starting from 90 ± 10°C.

TABLE 7.9
FMEA-like Worksheet for the Main FBG Components Involved in the Analysis

ID and Component	Component Function	Failure Mode	Causes	Frequences [1/h]
TIC101/TI101 Temperature Controller and Indicator	Temperature control (given the setpoint and the relative tolerance).	**FAIL negative**: inlet air temperature will try to be diminished.	PLC failure OR Transmitter Failure (a higher air inlet temperature is wrongly detected)	1.31E-06 (∑) (EXIDA)
		FAIL positive: inlet air temperature will try to be increased.	PLC failure OR Transmitter Failure (a lower air inlet temperature is wrongly detected)	1.31E-06 (∑) (EXIDA)
		Fail No signal: blockage on the previous signal (hp: FAIL positive)	PLC failure OR Transmitter Failure (no measure)	1.31E-06 (∑) (EXIDA)
V101 Solenoid Valve	Opening and Closure of the hot vapors for air heating	Fail open/close	Electric Failure Mechanical failure of one or more components	3.00E-05 [31]
TAH201 Alarm	Alarm of High Product Temperature	Failure	Electric Failure Mechanical failure of one or more components	2.90E-03 [1/y] [31]
X201/X202	Venting panel	Fail to open	Mechanical failure	1.00E-03 [1/y] (assumed)
GS201	Grounding System	Fail to ground	Electric Failure	1.00E-04 [1/y] (assumed)
KC101 Face & bypass damper	Air inlet mixing (and conditioning)	**FAIL to open (heating line)**	Electric Failure Mechanical failure	6.00E-06 [31]
		FAIL to close (heating line)	Electric Failure Mechanical failure	6.00E-06 [31]
		FAIL to operate (both open/close)	Wearing	6.00E-06 [31]
Human being	**LOAD HE** (Human error in API's loading)	Error in performing a given procedure	(not specified)	2.00E-03 [1/y] (assumed)
S200	General switch	Fail to shut off the plant	Electric Failure	1.00E-03 [1/y] (assumed)

No information concerning MIT, LIT, and MEC are available. Data available about the powder suggest a moderate approach regarding safety issues without proceedings with further experimental investigations before the ROA.

Finally, it is necessary to list all the main process variables to be analyzed in the ROA table. Table 7.10 reports all the considered variables.

7.3.4 Recursive Operability Analysis, Fault Tree Compiling, and Risk Evaluation

The application of the Recursive Operability Analysis (ROA), presented in the following, is based on a set of simplifying assumptions:

- Given the type of components and the overall plant configuration, common-cause failures were not considered in the evaluation of the accident-event frequency.
- Spurious failures were excluded from the analysis, as the components involved exhibit very high reliability with respect to this failure mode.
- The acceptable order of magnitude of the Top Event unavailability is 1.00E-06 (with a mission time of 1 year). Any result significantly deviating from this order of magnitude cannot be considered acceptable from a safety point of view.
- Catastrophic events, such as meteorite impacts, aircraft crashes, or natural disasters, were not considered as initiating events.
- Incidental phenomena, including fires or explosions originating outside the system under analysis, were also excluded as initiating events.
- Lightning strikes were excluded as a potential ignition source because the FBG unit effectively behaves as a Faraday cage; moreover, the plant is not operated during thunderstorms.

TABLE 7.10
Variables to Be Analyzed in the ROA Table

Symbol	Variable Deviation
F	Inlet Air Flowrate
T_{IN}	Inlet Air Temperature
T_m	Metal Grid Temperature
T	Product Temperature
m	API Load
P	Pressure
C	Product Concentration

- Flame flashback from the bin unit to the FBG unit is excluded, even in the event of combustion of fine dust accumulated, since such a scenario would not be capable of generating a spark with energy exceeding 500 mJ.
- Electrostatic charge accumulation during pneumatic conveying was excluded.
- Ignition sources, capable of generating sparks with energy equal to or greater than the minimum ignition energy of the API (500 mJ), were considered not credible.

Top Events of interest that do not lead to an explosive scenario (TE3) can include (a) formation of a dust cloud with concentration during fluidization major with respect to the MEC of API (TE1); (b) "caramelization" (TE2), i.e., polymerization following the opening of the epoxide ring (excluding concurrent conditions that could lead to an explosion) on the metal grid (N201); (c) ineffective granulation (TE4) due to malfunction of the binder spray system (water is used as the solvent in the FBG process). These events may have significant implications for product quality and process efficiency, but they do not result in an explosion hazard; therefore, only TE1 and TE2 were considered in the present analysis, while TE4 was excluded.

The evaluation of the expected frequencies of the identified accidental events can be carried out using a dual approach: construction (and mathematical resolution) of logical fault trees and use of reliability data obtained from internationally recognized specialist databases. The range of values adopted for failure frequencies or probabilities of non-intervention for each type of component (e.g., valves, dampers, controllers, etc.) was defined based on information provided by the main national and international reliability databases. The analysis does not take into account intentional acts of sabotage, nor incidents that may occur during maintenance activities while the plant is shut down. Based on the calculated frequencies of occurrence, the events can be classified as "probable," "improbable," or "remote or reasonably not credible." Within the scope of the present safety analysis, primary events are considered "reasonably not credible" when their frequency of occurrence is lower than 1.00E-09 events per year.

Table 7.11 reports an extract (that is only for TE1, TE2, and TE3) of the ROA table for the FBG unit.

After completion of the Recursive Operability Analysis table, OpenFTA software can be used to either represent or calculate the frequency of the selected Top Events, starting from the ISDs generated through the ROA methodology. The frequencies of the basic events are reported in Table 7.9. As already stated, in the present analysis, only three Top Events have been considered: formation of a dust cloud (TE1), caramelization of API on the metal grid (TE2), and dust explosion because of an internal ignition source (TE3), specifically either a hot surface or a spark generated within the FBG unit. External ignition sources and alternative explosion scenarios

TABLE 7.11
ROA Worksheet (Extract) for the FBG Unit

					Protections				
					Manual				
REC	NDV	Causes	Consequences due to Protection Failure	Plant State with Protections Working Correctly	Alarm (Optical/ Acoustic)	Operator Actions on Components	Automatic Protective Systems	Notes	TE
1	2.h.m (High Mass of API)	LOAD HE (Human Error in API Loading)	2.h.C (High Concentration> MEC)	2.h.C (High Concentration> MEC)					
2	2.h.C (High Concentration> MEC)	Rec 1	Dangerous Dust Cloud in the FBG	Dangerous Dust Cloud in the FBG	–	–	–	–	TE1
3	1.h.T_{IN} (High Inlet Temperature)	TIC101 Fail Positive OR KC101 Fail Close OR KC101 Fail to operate OR V101 Fail Open	2.h.T_m (High Metal Grid Temperature)	2.h.T_m (High Metal Grid Temperature)	–	–	–	–	–

4	2.h.T_m (High Metal Grid Temperature)	Rec 3	2.h.T (High Product Temperature) (Caramelization of API on the grid)	2.h.T (High Product Temperature) (Caramelization of API on the grid)	–	–	–	On the metallic grid exothermic reaction can occur, leading to a caramelization phenomenon.	TE2
5	2.h.T (High Product Temperature)	Rec 4	2.hh.T (Very High Product Temperature > MIT)	Caramelization of API on the grid	TAH201	Switch off of the plant	–		
6	2.hh.T (Very High Product Temperature > MIT)	2.h.T (High Product Temperature)	3.h.P (High Pressure)	3.h.P (High Pressure)	–	–	–		
7	3.h.P (High Pressure)	Rec 6	3.hh.P (Very High Pressure)	Plant Shut down and venting to atmosphere	–	–	X201/X202		
8	3.hh.P (Very High Pressure)	(Rec 7 AND 2.h.C) OR (GS201 Fail AND 2.h.C)	Explosion	Explosion					TE3

were deliberately excluded, in accordance with the assumptions previously defined in the ROA framework.

For the fault-tree representation of the Top Event "TE1: Formation of a dangerous dust cloud inside the FBG," it is necessary to refer to Rec 1 and 2 of Table 7.11. As it is possible to observe from Figure 7.11, TE1 unavoidably occurs whenever an API loading error occurs. No protective means are present; therefore, a dust cloud whose concentration can exceed the API's MEC can easily be reached. In this case, the unavailability of TE1 coincides with the unavailability of the corresponding human error (that is, an error in performing a given procedure). Considering that its failure rate, which can be found in Table 7.9, is 2.00E-03 [1/y]; it is possible to calculate an expected unavailability of TE1 equal to 2.00E-03 when a mission time of one year is considered.

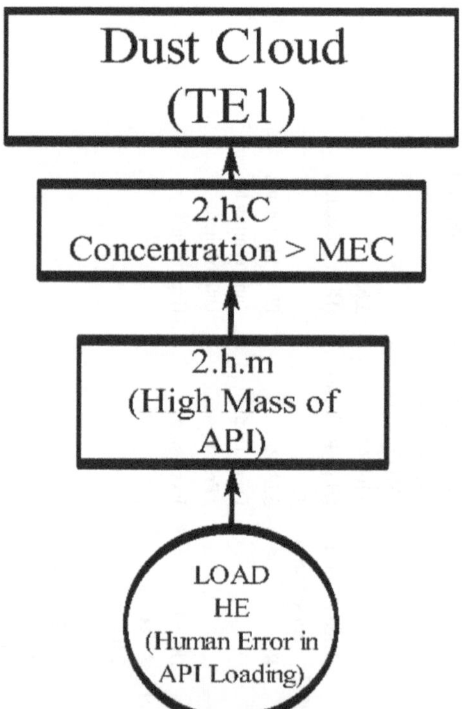

FIGURE 7.11 Fault tree representation for TE1.

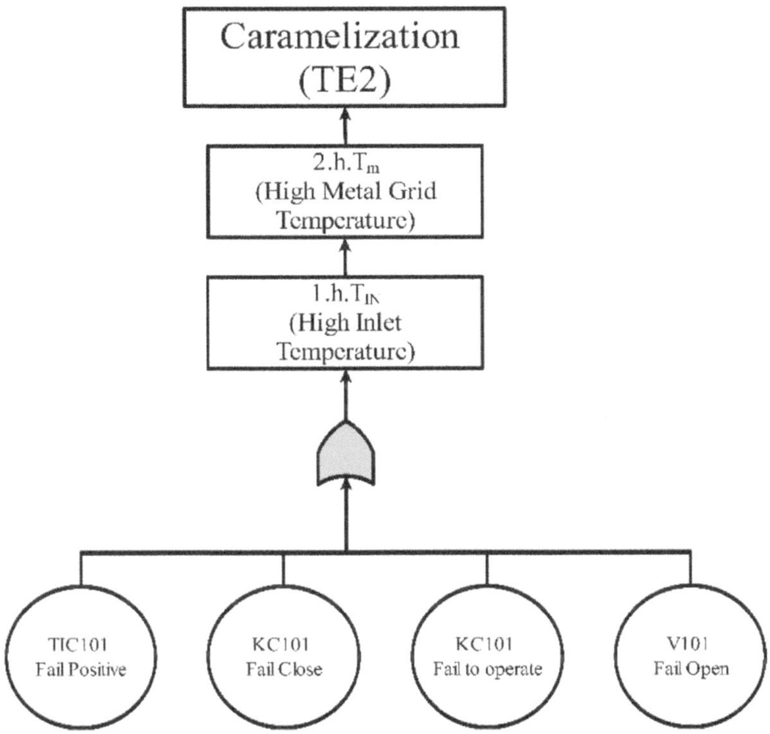

FIGURE 7.12 Fault tree representation for TE2.

The second TE, "TE2: caramelization of the API on the metal grid," is independent of TE1 and it can be represented as reported in Figure 7.12.

The numerical evaluation of this tree yielded an unavailability of 1.00E-01, always considering a mission time of one year. This value is very high because all MCSs are of first order.

Finally, the last TE, "TE3: dust explosion," is dependent on the other TEs, and it can be represented as reported in Figure 7.13.

OpenFTA analysis identified nine Minimal Cut Sets (MCSs), one of second order and the other of fifth order. The calculated unavailability of the Top Event, as directly obtained from the fault tree resolution performed by OpenFTA, is 2.00E-07. Such a value is lower than 1.00E-06, therefore, it can be considered fully acceptable and no further analysis (e.g., numerical simulation of the effects of the primary internal explosion of the FBG unit) are required.

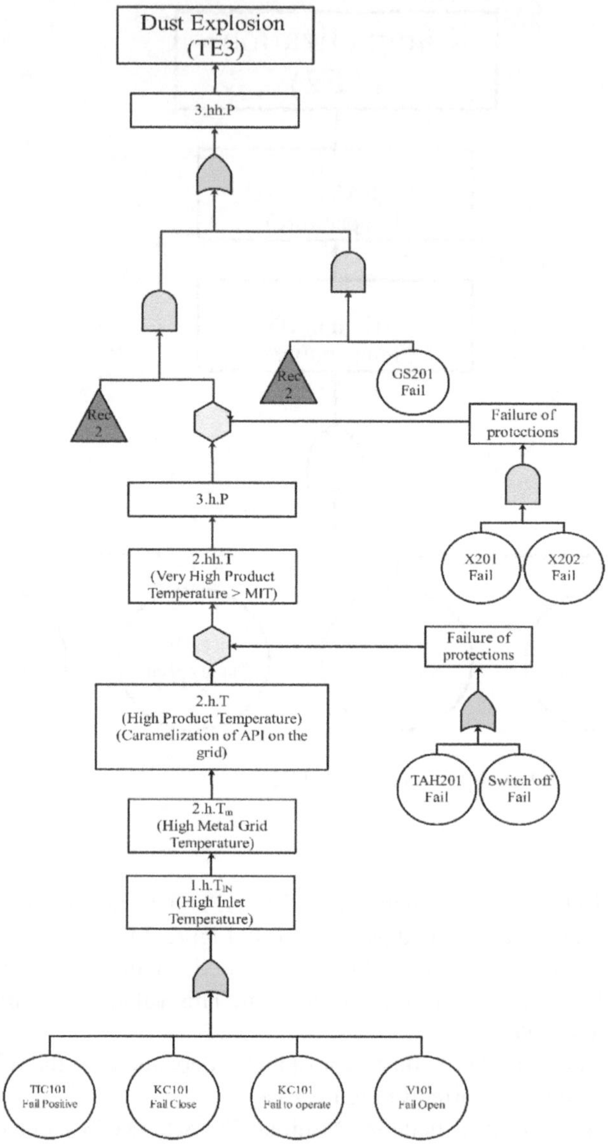

FIGURE 7.13 Fault tree representation for TE3: internal dust explosion of the FBG unit.

REFERENCES

1. Amin M.T., Khan F., Amyotte P., 2019. A bibliometric review of process safety and risk analysis. *Process Saf. Environ. Prot.*, 126, 366–381.
2. Mkpat E., Reniers G., Cozzani V., 2018. Process safety education: A literature review. *Journal of Loss Prevention in the Process Industries*, 54, 18–27.

3. Casal J., 2018. Chapter 11—Quantitative risk analysis. In *Evaluation of the Effects and Consequences of Major Accidents in Industrial Plants*, 2nd ed. Elsevier B.V. Amsterdam, The Netherlands, 439–481.
4. Goerlandt F., Khakzad N., Reniers G., 2017. Validity and validation of safety-related quantitative risk analysis: A review. *Risk Anal. Valid. Trust Risk Manag.*, 99, 127–139.
5. Smith D.J., 2017. *Reliability, Maintainability and Risk: Practical Methods for Engineers.* Butterworth-Heinemann, Oxford, UK.
6. Kletz T.A., 2001. *Hazop and Hazan: Identifying and Assessing Process Industry Hazards.* IChemE, London, UK.
7. Liu H.C., 2019. *Improved FMEA Methods for Proactive Healthcare Risk Analysis.* Springer, Singapore.
8. Kritzinger D., 2017. 4—Fault tree analysis. In *Aircraft System Safety.* Woodhead Publishing, Cambridge, UK, 59–99.
9. Ruijters E., Stoelinga M., 2015. Fault tree analysis: A survey of the state-of-the-art in modeling, analysis and tools. *Computer Science Review*, 15–16, 29–62.
10. Khan F.I., Abbasi S.A., 1998. MAXCRED—A new software package for rapid risk assessment in chemical process industries. *Environmental Modelling & Software*, 14, 11–25.
11. Khan F.I., Iqbal A., Ramesh N., Abbasi S.A., 2001. SCAP: A new methodology for safety management based on feedback from credible accident-probabilistic fault tree analysis system. *Journal of Hazardous Materials*, 87, 23–56.
12. Khakzad N., Khan F., Amyotte P., 2012. Dynamic risk analysis using bow-tie approach. *Reliability Engineering & System Safety*, 104, 36–44.
13. Piccinini N., Ciarambino I., 1997. Operability analysis devoted to the development of logic trees. *Reliability Engineering & System Safety*, 55, 227–241.
14. Piccinini N., Scarrone M., Ciarambino I., 1994. Probabilistic analysis of transient events by an event tree directly extracted from operability analysis: Evaluation of accidental releases of gas from a pressure-regulating installation. *Journal of Loss Prevention in the Process Industries*, 7, 23–32.
15. Arena U., Romeo E., Mastellone M.L., 2008. Recursive operability analysis of a pilot plant gasifier. *Journal of Loss Prevention in the Process Industries*, 21, 50–65.
16. Demichela M., Marmo L., Piccinini N., 2002. Recursive operability analysis of a complex plant with multiple protection devices. *Reliability Engineering & System Safety*, 77, 301–308.
17. Contini S., Contini P.M., Torretta V., Cattaneo C.S., Raboni M., Copelli S., 2016. Comparison of classical and "Cause-Consequence Diagrams" recursive operability analysis: The T2 laboratories accident. *Chemical Engineering Transactions*, 53, 109–114.
18. Contini P.M., Contini S., Copelli S., Rota R., Demichela M., 2015. From HazOp study to automatic construction of cause consequence diagrams for frequency calculation of hazardous plant states. In *Safety and Reliability of Complex Engineered Systems—Proceedings of the 25th European Safety and Reliability Conference*, ESREL 2015. CRC Press, Boca Raton, FL, 347–355.
19. Amyotte P., 2013. *An Introduction to Dust Explosions.* Butterworth-Heinemann, Boston, MA.
20. Davis S., Hinze P., Hansen O., van Wingerden K., 2011. Does your facility have a dust problem: Methods for evaluating dust explosion hazards. *Journal of Loss Prevention in the Process Industries*, 24, 837–846.
21. Eckhoff R.K., 2003. *Dust Explosions in the Process Industries.* Elsevier B.V., Amsterdam, The Netherlands.

22. Khan F.I., Abbasi S.A., 1999. PROFAT: A user friendly system for probabilistic fault tree analysis. *Process Safety Progress,* 18, 42–49.
23. Matuzas V., Contini S., 2012. *European Commission, Joint Research Centre, and Institute for the Protection and the Security of the Citizen. ASTRA Plus User Manual: Description of How to Use the Modules for Fault Tree Analysis and Concurrent Importance and Sensitivity Analysis.* Publications Office, Luxembourg.
24. Abuswer M., Amyotte P., Khan F., 2013. A quantitative risk management framework for dust and hybrid mixture explosions. *Journal of Loss Prevention in the Process Industries*, 26, 283–289.
25. Eckhoff R.K., 2019. Measurement of minimum ignition energies (MIEs) of dust clouds—History, present, future. *Journal of Loss Prevention in the Process Industries*, 61, 147–159.
26. Fumagalli A., Derudi M., Rota R., Copelli S., 2016. Estimation of the deflagration index K_{St} for dust explosions: A review. *Journal of Loss Prevention in the Process Industries*, 44, 311–322.
27. Kletz T., Amyotte P., 2019. Chapter 7—Accidents said to be due to human error. In *What Went Wrong?* 6th ed. Elsevier B.V., Butterworth-Heinemann, Boston, MA, 153–172.
28. Bello G.C., Colombari V., 1980. The human factors in risk analyses of process plants: The control room operator model 'TESEO'. *Reliability Engineering*, 1, 3–14.
29. Marmo L., Crivelletto V., Starace A., 2009. Recursive operability analysis as a decision support tool for risk-based maintenance. *Journal of Loss Prevention in the Process Industries*, 22, 557–565.
30. Imperial Sugar Company Dust Explosion and Fire | CSB. Available online: https://www.csb.gov/imperialsugar-company-dust-explosion-and-fire/, accessed on 01/11/2026.
31. Lees F., 2005. *Loss Prevention in the Process Industries: Hazard Identification, Assessment and Control.* Elsevier B.V., Amsterdam, The Netherlands.
32. Bundaber Sugar Ltd. Bundaberg Icing Sugar MSDS No. 4.1.1.1. 11 June 2012. Available online: https: //www.bundysugar.com.au/files/MSDS%20Icing%20Mixture_0.pdf, accessed on 01/12/2026.
33. Copelli S., Barozzi M., Scotton M.S., Fumagalli A., Derudi M., Rota R., 2019. A predictive model for the estimation of the deflagration index of organic dusts. *Process Safety and Environmental Protection*, 126, 329–338.
34. Jones-Lee M., Aven T., 2011. ALARP — What does it really mean? *Reliability Engineering & System Safety*, 96, 877–882.

8 Future Perspectives on Dust Explosion Safety

8.1 SUGGESTIONS FOR STANDARDS IMPLEMENTATION AND DEVELOPMENT

The importance of updating and progressively integrating standards can be illustrated by the recent reorganization of US regulations on dust explosions, as discussed in Chapter 5, which led to the creation of NFPA 660. This can be viewed as a process of optimization of existing regulatory resources, with the aim of making both their applicability and their usability more efficient, through easier retrieval of information and specific intervention procedures. The intention underlying this process shares the purpose of the present chapter, which seeks to make its modest contribution to stimulating constructive criticism of existing standards, with a view toward the development of new standards where necessary.

8.1.1 European Standards: Analysis for Implementation and Development

This section focuses on a critical analysis of the European Union regulatory framework for dust explosion safety in industrial settings. The aim is to provide food for thought on certain elements and stimulate the ever-growing awareness of the topic.

8.1.1.1 Safety Data Sheet (SdS): Standardize Dust Explosion Key Parameters

The compilation of product safety data sheets is, at present, a process described and defined within Regulation (EC) No. 1907/2006 (REACH), in Annex II "Guide to the compilation of safety data sheets," as amended by Regulation (EU) 2020/878 and each substance recognized as hazardous, including certain dusts, should, therefore, be accompanied by its own duly completed SDS. However, it is not uncommon to encounter combustible dusts whose SDS do not report any information on flammability or explosivity. In this regard, it is interesting to refer to an analysis conducted by Barozzi et al. (2025). The aim of this study was to assess the actual availability of dust explosivity data in different types of companies, and for this reason, 19 safety reports issued between 2021 and 2023, relating to 213 different types of combustible dusts, were examined. Considering parameters such as particle size, MIE, MEC, LOC, MIT, LIT, K_{st}, P_{max}, resistivity, and Combustion Index (CI), it was possible to estimate how, with respect to each individual parameter, the actual availability of data varied according to the following percentages:

DOI: 10.1201/9781003645320-8

- 0%–3% of the data were obtained from *ad hoc* tests.
- 0%–8% of data were available internally to the companies for tests already performed.
- 10%–15% of data were obtained from bibliography.
- 45%–85% of the data were missing.

The possible causes that have led to this significant lack of data (from 45% for MIE to 85% for CI) are: insufficient safety culture related to dust explosions, the time and money-consuming nature of detailed dust analyses and inappropriate compilation of SDSs. The REACH Regulation, through Annex II and subsequent amendments and integrations, provides that the 16 characteristic sections of each safety data sheet contain information useful for defining the measures necessary to protect the health and safety of workers and the environment, taking into account the potential intrinsic risks of individual substances. Despite this, it is possible that for a dust known to be combustible, but for which specific explosivity data are not available (e.g., K_{st}, MIE, MIT, …), it may be considered sufficient to warn the user of the potential risk of explosion, effectively delegating to the latter part of the investigation should it be necessary to obtain explosivity data. The REACH Regulation and its amendments further specify that the compilation must be carried out by a "competent person," adequately kept up to date, without, however, providing any detail on the specific requirements or competencies that this person must possess, thus leaving a wide margin of flexibility and subjectivity in the choice of who will be responsible for drafting the safety data sheets.

In fact, what emerges from the present analysis is the absence of specific obligations concerning the collection and sharing of dust explosivity data within SDSs, which results in a lack of uniformity in the quality and completeness of the information available to users. Structuring guidelines or a regulatory proposal capable of standardizing and harmonizing, at the European level, the minimum information and data that must be reported in safety data sheets regarding dust explosivity could prove to be a useful and necessary tool both to enhance the sharing of fundamental and important information and, consequently, to strengthen the prevention of accidental events. Such a proposal could also be integrated with a more objective and specific description of the technical and experiential characteristics that the person responsible for drafting safety data sheets should possess. Moreover, it should be structured while keeping in mind the practical and applicative needs required of it, avoiding the creation of further unnecessary bureaucratic complications. Defining the choice of sources from which to derive explosivity data (laboratory tests, bibliographic sources, …), the type of data required, applicative constraints and case reviews, and the ways in which it is possible to be justifiably exempt from reporting these data, in addition to the description of other parameters necessary to make such a proposal robust and concretely applicable, falls outside the scope of this section, which instead merely seeks to stimulate and support constructive debate for the implementation of corporate safety.

8.1.1.2 DHA: Introduction in EU Legislation

At present, European legislation does not provide for any type of obligation to carry out DHA for sites and facilities that handle combustible dusts. In this regulatory scenario, therefore, professionals engaged by companies to perform this activity are required to refer to US standards, such as NFPA 660, described in Chapter 5. The absence of a uniform standard applicable across the EU thus limits the ability of the industrial sector to protect itself against dust explosions and delegates to the managers of individual plants the responsibility of deciding whether or not to initiate a specific site characterization investigation with respect to this type of disastrous event, in line with US references. A brief comparison between European and US regulatory approaches is provided below to highlight the substantial difference in the implementation of DHA between the EU and the USA:

- In the EU, the approach to assessing the risk of dust explosions is regulated by the ATEX Directives (114 and 137) and, in part, by Directive 2012/18/EU (Seveso III), but it does not provide for an explicit obligation to carry out DHA analyses according to commonly shared criteria. The guidance provided is generally to demonstrate, through documentation, that the risk of dust explosion has been adequately assessed, primarily with regard to the protection of workers' safety.
- In the USA, reference can be made to the NFPA 660 standard, which in Chapter 7, sets out the obligations and references for carrying out DHA in facilities that use combustible dusts, also defining the frequency with which the analysis must be repeated (five years), with particular reference to process safety.

In synthesis, therefore, the lack of specific legislation that obliges companies, under certain specific conditions, to carry out DHA poses serious and concrete potential risks, such as:

1. *Different levels of safety among EU Member States.* Without a harmonized obligation, some States or sectors may apply stricter practices than others, resulting in a "mismatch" in industrial safety for companies operating in multiple countries. This can lead to more vulnerable facilities in less stringent jurisdictions, even at the same level of risk, with a consequent overall increase in risk across international supply chains. From an economic perspective, this could also be seen by less ethical companies as an opportunity to transfer higher-risk production to Member States (or even outside the EU) where controls are less stringent. Moreover, if an incident were to occur in a country with more lenient regulations, the entire European sector/brand could suffer repercussions.
2. *Complications in implementing effective technical measures to counter the risk of dust explosions.* Standardized and specific methodologies, such as DHA analysis, require specific skills and procedures and, at the same time,

are able to provide increasingly functional and refined tools to achieve the intended goal. Without a binding regulatory reference, many companies will not carry out a DHA intervention or will do so in an incomplete and ineffective manner.

3. *Greater complexity for inspection activities.* A lack of regulatory harmonization can complicate inspections and their comparison due to the possible and varying levels of safety defined by local regulations. The absence of harmonization would also make post-incident activities more difficult, such as comparing disastrous events with one another and preventing their recurrence under the same initial conditions.
4. *Possible slowdowns in insurance/commercial relationships.* At an international level, it is reasonable to expect that insurers and clients may require objective evidence, including that related to industrial safety (such as DHA analysis), before completing an insurance coverage process or product supply, also demonstrating a clear correlation between safety and profit. Furthermore, the marketing of products from the EU to the USA could be favored for European companies able to demonstrate that they have conducted DHA, precisely because it is a procedure widely recognized by process companies in the USA.

Building on the US experience, it is, therefore, possible to assume that the presence of a well-structured regulation for the adoption of procedures such as DHA would significantly contribute to a more targeted, systematic, and traceable management of combustible dusts used across various sectors of process industries, practically supporting the resolution of some issues related to its absence, including those mentioned in this section from 1 to 4.

A European proposal integrating the ATEX directives with requirements for DHA would significantly contribute to the development of industrial safety in every sector affected by the presence of combustible dust.

8.1.1.3 Dust Characterization: Mandatory Test or Mathematical Modeling

As a complement to the previous Section 8.1.1.1, and in the event of a possible regulatory requirement to report dust explosivity data in SDSs, it would seem appropriate to also define constraints regarding the execution of minimal laboratory tests for dust characterization in terms of explosivity. The intention is, therefore, to bring attention to the structuring of a possible regulation that would require manufacturers to test every combustible dust for which they issue an SDS, or to test them in cases where combustibility is unknown, so as to exclude or report, with objective data, the risk of combustion and explosion. The adoption of a consolidated and well-structured practice on this topic would allow us to:

a. *To classify a wide range of dusts with greater precision and specificity in relation to their explosivity.* This aspect would be highly important for reducing the risk of accidents, as companies would have access to specific explosivity data for the dusts they use in various industrial processes. The data employed by companies for analyses conducted for preventive purposes (e.g., DHA) would, therefore, derive from verified tests, since SDSs would no longer be compiled preferring literature data. The design of protection systems could also benefit from this practice, as the availability of more reliable data would allow for more accurate and appropriate sizing of ventilation systems, explosion protection, pressure relief, and other measures. On the other hand, such a procedure could be hindered by increased costs for manufacturers/importers, longer certification release times, and greater bureaucratic complexity, which any potential reference regulation should absolutely take into account.

b. *Stimulate the certification/accreditation process and continuous improvement of the testing laboratories involved.* The implications of this process, driven by the need for reliable values of the main dust explosive parameters, would have repercussions on the entire supply chain involved, as:
    1. It would increase the reliability and reproducibility of test results, reducing the risk of unreliable or misleading results. This is a crucial added value, especially for limiting parameters (e.g., MIE, K_{st}, etc.), especially for new types of dust.
    2. It would standardize testing methods and procedures across all laboratories, producing a general increase in the quality of individual laboratories.
    3. Continuous training for operators and the search for new sectors and products would be encouraged, opening the doors to new markets.

Finally, considering that the aforementioned notes are closely related to the issuance of SDSs, it would also be important to take into account the dusts present in industries for which the issuance of an SDS is not required. In such cases, the possible need to carry out laboratory tests for the characterization of the individual product should be evaluated, but it could also be possible to estimate these parameters using validated and reliable mathematical models.

8.1.1.4 Professionality: DHA e ATEX Expert Certification

As a complement to the previous Section 8.1.1.2, a note is proposed regarding the possible definition of professional figures specialized in dust explosion matters that are recognized at the European level. At present, there are no shared EU criteria for the qualification and certification of specialists capable of performing DHA and specific ATEX activities, sector inspections, and critically analyzing the results of analyses and tests, as well as designing protection systems against dust explosions. It should be acknowledged that private courses (not always recognized at a supranational level) and national identifications (e.g., Gefahrstoffbeauftragter in Germany, as hazardous substances officer) do exist, but, as they are not harmonized among themselves, they contribute only partially and in an unstructured way to the creation of a certified professional operating across the EU. The establishment of a European certification system for "Dust Safety Experts – EU certified" with structured periodic updates would compensate for an EU regulatory gap and contribute to standardizing the technical competencies of experts at the European level, while simultaneously enhancing the quality of European industrial safety.

8.1.2 US Standards: Analysis for Implementation and Development

The USA has recently undergone a significant reorganization of regulations on dust explosions, as already reported in Chapter 5. The NFPA 660 standard is notable for representing a considerable evolution in the regulatory field of dust explosions, as it unifies, clarifies, and makes the management of this specific industrial risk more systematic. The aim, as previously mentioned, is to facilitate the adoption of regulations, which, in light of this update, are condensed into a single reference document. Ambiguities are reduced, contradictions are eliminated, and those responsible for managing a facility now have a single reference source to consult. The conduct of DHA activities is also facilitated, as a reference guideline is made available, and the importance of regularly performing this preventive investigation is emphasized. This represents a revolution in the sector, allowing an enrichment in terms of corporate safety against dust explosions and, as indicated by Cloney C. [1]:

> *As industries embrace the one dust standard with NFPA 660 and prioritize dust hazard safety, the prospect of reducing the incidence of fires and explosions becomes increasingly attainable in existing facilities. Through collective action, public comment, and commitment to safety excellence, NFPA 660 paves the way for a future where combustible dust hazards are mitigated, with combustible dust standards, and workplace safety is paramount.*

Over time, and with the observation of the applicability of the NFPA 660 standard, it will be easier to highlight its strengths and weaknesses, thereby identifying aspects on which it would be appropriate to intervene to further enhance industrial safety in the USA.

8.1.3 Others: Analysis for Implementation and Development

The analysis conducted regarding possible implementations of applicable industrial safety regulations has been focused on the USA and the EU, but it could be interesting to attempt a similar generic analysis for other countries, particularly affected by the phenomenon of dust explosions. It would, therefore, be appropriate to observe the conditions affecting regions of the world to which a large part of the global industry, both process and manufacturing, is continuing to relocate.

In this regard, a brief and as careful as possible analysis will be presented concerning two giants of global industrial production: China and India. Market analyses related to these two countries describe and foresee increases in industrial production in the food processing and pharmaceutical sectors, which also draw attention to the possible repercussions that such increases could have on industrial accidents. In an analysis conducted by Maniar et al. (2020), for example, the incidental events affecting the global pharmaceutical industry between 1985 and 2019 were examined, and it was noted that of the 75 Process Safety incidents recorded, 83% occurred in China and India, 71% were explosions (of 51 recorded explosions, 10 were caused by dust), and 89% resulted in fatalities [2]. In relation to this important aspect, it is also necessary to examine in depth the trajectories that the two countries are following to build their industrial future.

- China is continuing its evolution from a "low-cost factory of the world" to a society capable of exporting high-quality products. In this scenario, the implementation of appropriate industrial safety systems could also make an important contribution to the ongoing industrial growth [3].
- The incentive plan defined by the Indian government (PLI, Production Linked Incentive) in favor of industries capable of implementing domestic production and export in various sectors (including pharmaceuticals) potentially has the capacity to significantly stimulate industrial growth in general, especially if the gains were reinvested in logistics, technology, and safety [4]. India is already considered the "pharmacy of the world," and Indian industrial development aims, in perspective, to create a pharmaceutical market valued at 130 billion dollars, more than double the current one, by 2030 [5].

The enormous industrial development potential foreseen for China and India, and the growth prospects in the global markets that concern them, can only be economically and socially interesting for the two Asian giants. However, in the general interest of workers, the environment, and the economy, these two countries will need to maintain active and dynamic investment in effective process safety, learning from the tragic accidental experiences of the past and preserving a perspective of continuous improvement.

8.1.3.1 China: Brief Analysis of Process Safety Evolution

Chinese industry currently maintains a state of developed industrial society which it has been able to build especially in recent decades [6]. Since the 1980s, China has experienced rapid and consistent growth from both an industrial and economic perspective. This situation has, therefore, created the inevitable need to consistently present itself as a competitive player in the global market and, at the same time, to structure a system of protection valid for worker safety. The relationship that generally links these two elements, and which deserves particular attention, is described by Liu et al. (2005) [7] and consists in the correlation between the peaks of Chinese economic development (1960, 1978, 1990, and 2003) and the increase in fatal accidents in the industry [8].

During its experience as a major industrial country, China has undergone not only significant industrial growth but also a necessary and progressive development of process safety, which, for simplicity, can be divided into three fundamental periods:

- Initial stage (before 2005)
  The introduction of the concept of Process Safety in China occurred in the 1990s, together with the arrival of foreign chemical companies that began opening facilities in Chinese territory. However, the true meaning of the term "Process Safety," both terminologically and practically, remained unclear until around 2006, as it was common to confuse "occupational safety" with "Process Safety," and only a few forward-thinking individuals, including some entrepreneurs, academics, and government personnel, began to inform themselves and train on this topic, which was progressively starting to enter the Chinese work approach.
- Start-up phase (between 2006 and 2013)
  Starting in 2006, the State Administration of Work Safety (SAWS) launched a series of important targeted programs and activities, including effective exchanges with international companies operating in China, aimed at implementing occupational safety management through the development of procedures and management plans for hazardous chemicals. During this significant period for Process Safety in China, the CCPS-CS served as a reference point for the development of this approach, and in 2010, the first non-mandatory standard related to process safety was issued called "Guidelines for Implementing Process Safety Management in Chemical Enterprises." At this point, the concept of Process Safety began to circulate more familiarly within Chinese industries.
- Rapid development phase (after 2013)
  After 2013, the Chinese government decided to massively promote the development and application of Process Safety in companies under its jurisdiction, which led to a rapid increase in the use of this approach across various production sectors. Greater dissemination of management systems, incentives for training and updates in safety matters, and increased use of tools such as HazOp, LOPA, SIL, and QRA are just some of the positive consequences of this significant governmental push [9].

Despite the efforts made by the entire Chinese industrial and governmental system to maximize its capacity to protect workers and ensure their safety, several significant and varied incidental events have occurred. Between 2005 and 2020, China experienced 67 officially recorded dust explosions. These events resulted in a total of 547 deaths and 600 injuries [10]. The development of an increasingly strong awareness and focus on industrial safety issues, together with the availability of increasingly sophisticated tools, allow even higher levels of protection for workers, the environment, and processes, especially considering that, after the significant industrial development experienced in recent decades, China can still be considered "the world's factory," with the implications this entails in terms of safety [11]. It is, therefore, reasonable and desirable to believe that, given China's role in the global production landscape, the Chinese government will maintain as priority occupational safety and will be able to guide the country toward continuous and progressive improvement of Process Safety practices in order to minimize the occurrence of incidents such as those caused by dust explosions.

8.1.3.2 India: Brief Analysis of Process Safety Evolution

Already from the early 2000s, it was possible to observe that India was generally one of the world's fastest-growing economies, and this condition was supported and stimulated by initiatives introduced by the Indian government to encourage investment by foreign companies. Industrial development in the territory was also favored by the so-called "China plus one" strategy, that is, the intention of large companies to have an alternative to the Chinese supply chain [12]. In relation to the focus of the present text, it is also essential to highlight that the Indian food and pharmaceutical industries are considered pillars of the Indian national economy [13, 14]

The Indian food processing sector is in a phase of significant growth and is projected to reach a market value of 535 billion US dollars by 2025–2026, thanks mainly to the increase in domestic consumption and exports [16], demonstrating its role as an economic driver for the entire country and becoming a priority for the Indian government. Despite its significant importance, the sector encounters some structural and technological challenges, including opportunities to further strengthen attention to quality and safety standards [13].

The Indian pharmaceutical manufacturing sector, in 2023, was ranked as the third largest in the world in terms of production volumes, and both the capacity and quality of the Indian pharmaceutical sector have been able to enter highly selective and regulated markets such as those of the USA, the United Kingdom, and the European Community [14], to which mainly generic drugs, active pharmaceutical ingredients, and vaccines are exported [17]. Furthermore, as already introduced in Section 2.1.2 of the present text, with a current economic value of around 50 billion US dollars, the future projection for this sector is to reach a market value of approximately 120 billion US dollars by 2030 and 450 billion US dollars by 2047 [14].

The evolutionary trajectory of the Indian process industry, particularly in the food and pharmaceutical sectors, inevitably entails, as in the Chinese situation, a growing concern regarding process safety and, specifically, the risk of dust explosions. At the conclusion of an analysis conducted by Vijayaraghavan [15], on the relationship between the phenomenon of dust explosions and the corresponding concern of the Indian industry, there seems to be "*Surprisingly very little awareness and almost no information exists on dust explosion in India. This makes it imperative that efforts are vested towards generating awareness and stimulating R&D on dust explosion in India.*" To date, based on some recent incidental events [18, 19], it can be stated that the issue of dust explosions in Indian process industries remains a topic that requires greater attention. The pharmaceutical sector was affected in July 2025 by yet another dust explosion, which, according to available information, appears to have been favored by the lack of appropriate analyses (DHA) on the risk of dust explosions [20]. The event caused the death of more than 40 workers, and, quoting the article, confirm that safety conditions for workers in India's industrial sector remain an area of ongoing concern [21]. Furthermore, the cross-cutting nature of process safety covers production plants of various sizes, which consequently have different structures and internal dynamics. Among these, in the Indian production sector, the role of well-known multinational pharmaceutical companies is significant, which sometimes, through unclear or incomplete management and sharing of safety data and information, can contribute to the occurrence of dramatic incidental events [22].

Similarly to the Chinese situation, it is, therefore, desirable that a continue joint effort be made between the entire Indian industrial sector, multinational companies, and the central government in an attempt to further improve the quality of occupational safety and Process Safety, so as to allow, in parallel with the increase in production volumes, greater guarantees in terms of protection for workers and the environment, greater work continuity and reduced economic losses in the interest of the entire Indian production sector.

8.2 EMERGING TECHNOLOGIES AND RESEARCH DIRECTIONS

In the analysis of emerging technologies, one must certainly consider the one that more than any other is defining the contemporary era: artificial intelligence (AI), understood as the development of algorithms and computer systems based on artificial neural networks that allow the interpretation of various data sources from which outputs, information, or models can be extracted. The widespread presence of AI (including its derivatives such as Machine Learning (ML) and the Internet of Things (IoT)) in research, production, financial, and domestic activities compels us to view it as the starting point of a new technological era and as a tool with a strong economic impact on a global scale. In this discussion, it is also necessary to underline that the ethical and moral implications related to the existence and use of these tools, along with further sector-specific considerations on the topic, go beyond the scope of the present text.

Classifying AI as "emerging" is paradoxical because its existence, in the form we know today, is the result of developments and research spanning several decades; however, it can still be considered as such in terms of the applications attributed to it today. The potential and development margins of AI are such that they impose the need to thoroughly understand this technology to use it to improve industrial activities, product quality, and various other areas of collective interest. Among these, the application of AI in the scientific field has influenced the work of researchers, scientists, and engineers, providing them with a powerful tool for the development of methods, devices, and systems capable of optimizing work processes, environmental protection activities, energy consumption, and many other relevant aspects.

However, it is necessary to remember that AI is nothing more than a tool, with the limitations that this imposes. Its development derives from a continuous learning process based on enormous amounts of data available on the internet, and among these, it is important to note that not only verified and reliable data exist, but also questionable data that could compromise the quality of AI learning. In these simple terms, necessarily without further detail, it is possible to expect that AI, when faced with complex or important decisions, may make the wrong choice. From an application standpoint in the field of occupational and process safety, the development of AI models aimed at resolving complex and important situations should, therefore, be considered only a partial solution to the problem, which would thus be appropriately addressed under the professional supervision of humans.

8.2.1 Toward Safety 5.0: Process Safety and AI

The transition from the concept of Industry 4.0 to the next level, known as Industry 5.0, is already underway. This fifth industrial phase aims to develop the relationship between humans and digital tools, which are the cornerstones of the Industry 4.0 model, toward a deeper integration in which this union can be considered a consolidated phenomenon and where the human-centric approach places the individual and their well-being at the center. The objectives of this transition focus primarily on process optimization, the implementation of sustainability across the entire industrial system, and the establishment of a flexible and resilient work model. All this is also possible, thanks to the important steps forward made in the field of AI, which is made available to man as a tool capable of supporting the development of the industrial system in achieving important targets, including environmental sustainability through the optimization of work processes and the reduction of energy consumption and, worker safety through the implementation of prevention and protection activities and tools in the workplace [23].

Process industries have been the activities most affected by the impact that these intelligent tools have brought to the production sector. Examples of real applications of AI in process industries stem from the presence of intelligent systems and models for proactive monitoring and prediction of equipment or plant failures, which allow for reduced downtime and optimized maintenance procedures, thanks to the intrinsic ability to operate in real time and make structured decisions based on the specific case observed [24]. AI has also found specific applications in the field of industrial safety, where it can be employed to recognize and protect against potential hazards, contributing to the improvement of prevention and protection measures [25]. However, the availability of these technologies, closely connected with industrial safety systems, presents companies with the serious necessity of considering other types of risks as well, such as cybersecurity, where potential attacks on automated systems can endanger the health and safety of workers and the environment [26]. The implementation of industrial collaboration between humans and intelligent digital tools has thus triggered the emergence of Safety 5.0, whose aim is to integrate the contribution of AI, digital technologies, and real-time analysis with the human component in the activities of risk analysis, management, and control. In fact, Safety 5.0 is, therefore, a fundamental element for the full development of the growth process toward Industry 5.0, in which safety will no longer be merely a priority, but an integral part of work activities [27].

8.2.2 Toward Safety 5.0: Dust Explosion and AI

The complexity of the phenomenon of dust explosions and the devastating consequences that can occur as a result of such events suggest that any efficient and valid contribution can be useful in preventing and mitigating these occurrences. In this regard, the areas and methods through which the Internet of Things (IoT), Machine Learning (ML), and, more generally, AI can support the development of specific prevention and protection tools and methods will be analyzed. Although they can be understood as separate entities, the joint and integrated work of IoT, ML, and AI enables the development of intelligent, proactive, and adaptive Safety Management Systems capable of reducing the contribution of human error and, in general, contributing to the improvement of the entire process industry sector [28].

8.2.2.1 Internet of Things (IoT)

The term Internet of Things (IoT) is conventionally associated with a virtual system in which interconnected digital and mechanical tools are capable of collecting, exchanging, and processing data in order to produce an output useful to the system itself. This new space, therefore, is configured as something dedicated to the relationship between physical objects, increasingly technologically advanced, that exchange and process information with human users in order to perform actions aimed at achieving a goal. Here too, an evolution toward a new digital environment can be observed, in which not only user-managed terminals will be connected, but also, increasingly, smart electronic devices [29].

The availability of IoT has been appreciated by the industrial and process sectors, which have recognized its potential to enhance performance by acquiring, using, and transmitting data at low costs while managing intra- and inter-organizational connectivity in a simplified manner [30]. In process industries, especially in complex plants, it is common to observe work activities in which the exchange of information and data occurs at multiple levels, involving both human users and smart devices in a flow that can be defined as complex. The inclusion of IoT in industrial processes, defined as IIoT (Industrial Internet of Things), therefore, represents a step in the digital transformation of certain operational stages toward a more efficient operational and management level, capable of accelerating decision-making activities and response times, with significant repercussions in various areas, including safety.

The direct application of the IIoT environment in process industries could, for example, allow the connection and rapid exchange of information between various plant elements such as sensors, actuators, IIoT gateways, plant units, smartphones, tablets, and other devices. This generates a data flow between devices that is managed almost automatically and can be recorded in real time for studies, analyses, and research in the field of Process Safety [31]. A fundamental aspect of introducing IIoT into work environments is the significant contribution it can make in reducing risk detection uncertainty. Through a more dynamic approach, in which work processes can be monitored in real time, it is possible to adjust process and safety parameters quickly, optimizing corrective interventions [32].

The potential of IIoT systems can, therefore, be exploited for specific industrial risk mitigation activities, including those related to dust explosions. Examples of implementations of prevention and protection measures against this type of incident include combustible dust explosion risk assessment software and models and software for the assessment of dust explosion risks and consequences (when integrated into an IIoT environment, it allows for greater capability in identifying critical issues, evaluating and quantifying potential explosion hazards, and developing plant monitoring procedures compared to isolated models). The main reasons supporting this implementation are the availability of different types of real-time data and the rapid sharing of information from connected smart devices [33].

Dust monitoring devices, for example, are designed to detect the presence of suspended dust material. The availability of a connection within an IIoT environment would allow these instruments to perform real-time monitoring and send rapid alerts in case of dust accumulation above the risk threshold, promoting swift intervention. In this way, these systems could be equipped with a proactive, fast tool capable of ensuring greater functionality and system reliability [34, 35].

The presence of ventilation and control systems in industrial buildings, however, is becoming increasingly common, and their development toward more efficient and smarter functionality is now seen as necessary, both in terms of energy optimization and human health. The application of IIoT technology to systems in this field would facilitate not only the achievement of these goals but also the implementation of industrial safety, as the exchange of data between sensors inside and outside the building would allow for effective dynamic regulation of airflow and recording for subsequent analysis [36]. In this scenario, it is, therefore, also possible to implement monitoring of internal dust in the work environment with specific sensors and automatically regulate the air intake, thereby keeping combustible dust concentration levels below the minimum risk threshold [33].

The continuous monitoring offered by IIoT systems allows for real-time analysis of process data and assessment of whether conditions are approaching a risk that must be avoided. In such cases, it is, therefore, possible to foreseen automatic activation of alarm and intervention systems in concerning situations, such as excessive dust accumulation, temperature increases above threshold levels, or reductions in environmental humidity that favor reaching explosive limit values.

Maintenance plays a key role in the proper functioning of machinery and plants and, consequently, in the correct execution of work activities. This preventive procedure also helps reduce plant downtime and maintenance costs, increase machinery lifespan, and improve safety conditions related to the work process. The use of IIoT technologies allows this practice to evolve from being preventive to predictive. Maintenance will no longer be seen as a standardized and scheduled practice to avoid plant stoppages or malfunctions, because predictive maintenance, enabled by the connection of intelligent devices within the IIoT environment, would allow the system itself to request intervention when necessary. In this way, it is also possible to anticipate future plant performance based on historical operating and maintenance data recorded over time. Especially for machinery that, due to improper maintenance, could generate hazardous conditions, it is necessary to implement maintenance with a proactive and dynamic strategy to prevent, among other things, explosions or dangerous dust releases in the workplace [37].

8.2.2.2 Machine Learning (ML)

Machine Learning (ML) is a branch of AI specialized in analyzing large amounts of data, recognizing trends and frequent behaviors of a system, and making predictions or decisions in a process aimed at continuous optimization. In practice, ML handles AI learning through the creation of algorithms capable of processing and combining available data in ways different from previous approaches, following a perspective of performance improvement. The initial datasets and the outputs generated by the system itself become the new input data that ML technology uses to learn and optimize the working algorithm.

The complexity of current industrial plants requires sophisticated mathematical tools capable of capturing the dynamics among the various involved equipment, and ML techniques could be able to perform this task. In relation to these potentials, ML has been rapidly adopted in process industries with the aim of innovating and making both existing processes and those related to the creation of new products more efficient [38]. The automatic learning pathway of digital systems through ML also allows for simplifying the optimization of process safety, for example, through its application in Predictive Maintenance (PM) activities as an integration with the capabilities of IIoT [39], improving performance in terms of data analysis, and predictive modeling [40].

Therefore, ML can be considered an evolutionary integration of IIoT, which, with this contribution, has at its disposal a tool capable of enhancing its analytical performance and, consequently, generating system outputs optimized with respect to process targets, including in terms of safety. In essence, while IIoT primarily has the task of collecting and sharing large amounts of data and generating an output based on the results obtained, the inclusion of ML in the process enables continuous system learning, whose outputs are used as new inputs to discover hidden patterns, develop working algorithms, make future predictions, and, therefore, improve the quality of output signals. In relation to this, it is possible to consider that all fields of application in which IIoT can be applied can also include ML.

An additional potential application of ML in Process Safety is related to its use in the analysis of incident reports. The study of the causes and consequences of incidents under similar conditions can provide a solid basis for guiding safety interventions and preparing detailed investigations. In this case, the use of ML makes it possible to automate and speed up a time-consuming task and, at the same time, to remove systematic errors (bias) capable of distorting the final results. Sattari et al. (2021) proposed an ML application method for the identification of incident causes through the analysis of incident reports. The different incidents were classified and used as inputs for a Bayesian Network capable of defining causal relationships among human factors, type of incident, and total number of incidents [41].

8.2.2.3 Artificial Intelligence (AI)

AI is generally capable of offering significant advantages in both dust explosion prevention and protection. Its effectiveness, however, depends on the reliability of the entire system it is integrated into. Therefore, this powerful tool must be integrated into a well-organized strategy capable of ensuring reliability at various levels.

Since AI can be seen as the container within which both IIoT and ML fall, in its generic nature, it can be included in all the practical applications of Process Safety implementation that have been briefly discussed for IIoT and ML.

The dust explosion prevention sector can benefit from the use of AI through:

- Continuous and real-time monitoring.
  The use of smart sensors interconnected with ML technology makes it possible to have a system capable of predicting risk conditions, activating dynamic alarm devices calibrated on actual operating conditions rather than on fixed parameters, and more rapidly sending intervention commands to prevent the attainment of threshold limit values associated with dust explosions.
- Instrumental data analysis and PM implementation.
  An advanced and intelligent management of plant data also makes it possible to implement PM procedures capable of predicting in advance the potential degradation of filters, cyclones, and other equipment involved in dust-related work processes. In this way, it is possible to reduce the likelihood of plant failures and malfunctions, with the consequent undesired dispersion of combustible dust or the generation of ignition sources [42].
- Estimation of dust explosive parameters.
  Specific AI models can be employed to optimize and speed up the tasks of data preparation and estimation of dust explosibility characterization values. Parameters such as MIE, K_{st}, or MEC can be estimated more easily based on the chemical–physical characteristics of the dust, reducing the number of tests that need to be carried out in the laboratory and simplifying the interpretation of the data obtained from the tests [43].
- Assisted simulations. Digital Twin e Management of Change (MOC).
  A possible evolution in the management of data originating from the IIoT environment is the use of Digital Twins (DT), a technology capable of virtually replicating objects, processes, or systems and updating them in real time as input data vary in order to simulate behaviors and performance. A DT can be used to address several critical aspects of industrial plants and processes, from design to the prevention and protection of major accidents. In plants affected by the presence of dust, this technique can be used to reproduce their operation and simulate various conditions that may lead to dust dispersion according to a "what if" approach. The simulation capabilities provided by DTs, in combination with advanced Computational Fluid Dynamics (CFD) models and ML technology, can also serve as support for optimizing the design of overpressure venting devices (vent panels). Plant DTs can also be used to assess the effects of operational changes and suggest corrective actions according to the Management of Change (MOC) approach [44].

The dust explosion protection sector, on the other hand, can benefit from the use of AI through:

- Early explosion detection with acoustic signals.
  The application of specific acoustic instrumentation, integrated with ML technology, can enable the recognition of the initial ignition phase of an explosion within fractions of a second. A potential application in the industrial field could allow the activation of suppression/isolation systems or shut-off valves [45].
- Automated checks and interventions.
  Based on the data shared within the IIoT environment, it is possible to set up an AI system capable of modulating the operational parameters of the plant or of specific instrumentation in order to optimize the process in terms of performance and safety, dynamically keeping individual operating parameters around their optimum and, at the same time, away from the minimum limits for the creation of conditions favorable to dust explosions.
- Integrated security systems.
  AI can be a valuable tool for developing integrated safety systems in which the operation of detection devices (e.g., acoustic detection or pressure variation monitoring), signaling systems (e.g., acoustic or visual), and control devices (e.g., valves and switches) can be interconnected in order to speed up human response and reaction times to an emergency condition [46].

8.2.3 Mathematical Tools for Dusts Explosivity Parameters Estimation

Each of us, when faced with complex situations, tends to adopt, more or less consciously, the same strategy: simplification. Scientists and engineers experience the same dynamic when they need to understand a complex system and make use of this problem-solving tool to create an understandable representation of the problem they must solve. The use of models is, therefore, more widespread than one might think, since, by definition, a model is nothing more than a simplified representation of a system [47]. The use of mathematical language as a symbolic system to express this simplification corresponds, in fact, to the use of a mathematical model to understand and solve the original system [48, 49]. Among the fields of application of mathematical models; therefore, there is also that relating to the study of dust explosions.

In this field, laboratory tests undoubtedly represent a fundamental tool for characterizing the explosiveness of powders, but conducting testing campaigns can prove to be lengthy and costly [50]. In the cases envisaged, however, the use of valid mathematical models proves to be a viable alternative to tests, especially for the estimation of MIE and K_{st} [51–53], which are the parameters most commonly used to characterize the explosiveness of a powder.

8.2.3.1 Mathematical Tools: Organic Dusts MIE Estimation

Copelli et al. (2021) [50] proposed a validated method for estimating the MIE value based on data from organic powders that can be easily obtained through thermogravimetric (TGA) and particle size analyses. The purpose of this tool is to mathematically simulate the ignition phenomenon occurring inside the Hartmann tube (or MIKE 3) in order to achieve an appreciable estimate of the MIE value of the analyzed powder. The key consideration of this model is precisely related to the mathematical representation of the ignition phenomenon, which is idealized as the transfer of energy from the electric arc, generated by the electrodes of the MIKE 3 tube, to the air–powder mixture suspended in the cylinder of the instrument. If the energy transfer from the electric arc is sufficient to heat the ignition region (hot core) of the air–powder mixture, the onset of pyrolysis could be hypothesized, and if the flammable volatile gases emitted fall within the flammability range, then an explosive phenomenon could be expected.

Once the scenario for the individual test in the Hartmann/MIKE tube is mathematically set, identifying the arc energy and the amount of powder, it is possible to proceed with the estimation of the equivalent diameter of the hot core and the air temperature. By then applying the model's equations, the concentration of volatiles ρ_v emitted by the powder can be obtained, and if this quantity falls within the flammability limits (within 120 ms, which is an arbitrary fixed value for the model), the ignition can be hypothesized. By iterating the procedure under different test conditions and applying the specific formula for identifying the MIE, see Equation (8.1), it is finally possible to accurately estimate this parameter.

$$MIE = 10^{\frac{[log10(E2) - I(E2)\cdot(log10(E2) - log10(E1))}{[(NI+I)(E2)+1]}} \tag{8.1}$$

E2 is the minimum energy at which at least one of the powder quantities ignites, E1 the maximum energy at which the ignition of all powder quantities always fails, I(E2) the number of powder quantities ignited at energy E2, and [NI + I](E2) the total number of tests (resulting in both powder ignition and non-ignition) performed at energy E2 [50].

An evolution of the previous model is developed by Copelli et al. 2025 [52], in which the estimation of the MIE parameter for organic powders is carried out through the introduction into the hot core of the concept of Critical Energy Density (CED [kJ/m³]), considered as the precursor approach to the LFL, briefly understood as the minimum energy required to sustain flame propagation. Equation (8.2) mathematically expresses the concept of CED.

$$CED = -\Delta H_{comb} \cdot (1-\beta) \cdot c_v \tag{8.2}$$

where ΔH_{comb} [kJ/mol] is the enthalpy of combustion of solid powder, β [-] represents the non-pyrolyzed solid fraction identified by tests TGA, and c_v [mol/m^3] is the molar concentration of volatiles in the hot core area.

The mathematical structure of this model is again based on data obtainable from rapid and cost-effective approaches, such as particle size analysis and thermogravimetric analysis combined with differential scanning calorimetry (TGA-DSC). The availability of data from these tests allows us to: (a) derive the pyrolysis kinetics of the considered organic powder; (b) estimate the effective enthalpy of pyrolysis; (c) deduce the effective enthalpy of combustion of volatiles emitted from the dust.

Using these data, it is, therefore, possible to calculate the CED value required for the specific test condition and to define in which scenarios the ignition or non-ignition of the powder is plausible. The modeling arises from the identification of a "hot air core," sized in terms of an equivalent sphere diameter (D_{HC}), defined as a function of the electric arc energy (E_S), as reported in Equation (8.3).

$$D_{HC} = 2.0993 \cdot E_S^{0.2871} \tag{8.3}$$

The air heated within the hot core by the electric arc allows the warming of powder particles and the consequent possible pyrolysis, and if the resulting mixture of flammable gases and air falls within the flammability limits (LFL–UFL), the triggering of combustion and the consequent explosion of the powder can be considered. The peculiarity of this specific model lies in the fact that the concept of LFL has been transposed into the concept of CED, required by the hot core to sustain the combustion of the mixture of combustible volatiles released from the powder inside it due to pyrolysis. The indicative CED value for most of the most common pyrolysis gases is around 1750 kJ/m^3, with the exception of H_2 (510 kJ/m^3) and CO (1376 kJ/m^3).

Once the set of tests in the Hartmann/MIKE tube has been simulated and the conditions in which the ignition or non-ignition of the powder can be reasonably assumed have been identified, it is possible to proceed with the estimation of the MIE using Equation (8.1). The estimated MIE values obtained from the model, when compared with those from actual laboratory tests, have shown remarkable reliability and accuracy, with a deviation of less than ±25%, allowing this model to be considered a valid tool for the analysis and assessment of powder explosion risks.

8.2.3.2 Mathematical Tools: Organic Dusts K_{st} Estimation

A second parameter of interest for mathematical models concerning powder explosions is the deflagration index K_{st}. Copelli et al. (2019) [51] introduced a validated model useful for estimating this parameter, related to the assessment of the magnitude of organic powder explosions, as if obtained from a test in a 20L sphere. Similarly to what was reported for the estimation of the MIE value, the possibility of using tools such as reliable predictive models for the evaluation of the K_{st} parameter allows, within the possible limits of use, companies to save time and money, especially in cases where it is necessary to test different powder usage conditions. The proposed model is valid for organic powders, and it is based on data obtainable from thermogravimetric testing, such as the estimation of the concentrated devolatilization kinetics.

The model is based on the assumption that within the testing apparatus, the action of the ignition device is slower than the combustion rate of the powder particles, which would lead to the formation of a fireball capable of enveloping the suspended powder before it begins the combustion process.

The mathematical framework of this tool is developed in the following sequential steps: (a) estimation of the thermal contribution provided by the chemical igniters; (b) characterization of the powder pyrolysis phenomenon and, finally, (c) characterization of the combustion of the air-combustible gas mixture.

The thermal contribution provided by the chemical igniters into the testing apparatus can be taken into account considering that an average power Q_{ig} is released after the triggering of the igniters themselves (the phenomenon is quite complex because it involves a metallic dust explosion).

Due to the heat provided by the igniters, a pyrolysis phenomenon implying the release of volatiles is triggered (of course, for organic dusts only). The pyrolysis phenomenon is mathematically modeled by assuming a single devolatilization step in which only a portion of the solid is converted into combustible volatile substances, while the remaining part forms the solid that remains in the char residue, called the skeleton. The equations of this "second block" will concern: mass balance for the solid, mass balance for the volatile substances (considered as a unique pseudo-species) and the energy balance for the solid particle. These equations can be found in the works of Copelli et al. [51].

Finally, the characterization of the combustion occurring in the gas phase can be done by writing both a material and an energy balance equation for the gaseous phase present into the sphere.

The material balance of the volatile substances within the sphere can be expressed by Equation (8.4).

$$\begin{cases} \dfrac{\partial \rho_V}{\partial t} = \dfrac{v\big|_{r=R} \cdot A_p \cdot N_p}{V_{sphere}} - k_{c,T} \cdot \rho_V \\ I.C. \qquad t = 0, \qquad \rho_V = 0 \end{cases} \tag{8.4}$$

where: ρ_V is the average concentration of volatiles in the gaseous phase, $v\big|_{r=R}$ is the rate of volatiles release in correspondence of the external surface of a single dust particle, A_p is the external surface of the particle, N_p is the estimated number of dust particles present into the hot core, V_{sphere} is the volume of the 20L sphere, and $k_{c,T}$ is the simplified kinetic constant associated with the combustion phenomenon of volatiles into the gas phase (which can be calculating by computing the characteristic combustion time, which is a function of the laminar burning velocity of the volatiles emitted during the pyrolysis phenomenon).

As it is possible to observe, the volatiles production rate ($v\big|_{r=R}$) is an essential information for calculating the average concentration of the volatiles themselves into the sphere gaseous phase.

The temperature of the air inside the sphere can be then estimated by writing the corresponding energy balance represented by Equation (8.5).

$$\begin{cases} \rho_{air} \cdot V_{sphere} \cdot c_{p,air} \cdot \dfrac{\partial T_{air}}{\partial t} = \Delta H_{com} \cdot k_{c,T} \cdot \rho_V \cdot V_{sphere} + h_c \cdot A_p \cdot N_p \cdot \left(T\big|_{r=R} - T_{air} \right) \\ I.C. \qquad t = 0, \qquad T_{air} = 293 \quad K \end{cases} \tag{8.5}$$

where ρ_{air} is the air density, V_{sphere} is the volume of the 20L sphere, $c_{p,air}$ is the specific heat capacity of the air inside the sphere, T_{air} is the temperature of the air inside the sphere, ΔH_{com} is the combustion enthalpy, h_c is the overall heat transfer coefficient, and $T|_{r=R}$ is the particle temperature in correspondence of its surface.

As it is possible to notice, Equation (8.5) neglects the radiative heat transfer from air to the particle, because it is very low if compared with the convective flux (that is, the heat exchanged between the air inside the sphere and the external surface of the powder particles).

Finally, the calculation of the deflagration index can be done, referring directly to the information regarding the trend of the temperature inside the sphere. In fact, solving the system of equations describing all the phenomena occurring during the explosive event, it is possible to compute the term $(dT/dt)_{max}$, which is the maximum rate of temperature rise during the phenomenon. K_{st} can then be expressed by Equation (8.6) [51].

$$K_{st} = \left(\frac{dP}{dt}\right)_{max} \cdot V_{sphere^{1/3}} = \left(\frac{n_{air} \cdot R}{V_{sphere}}\right) \cdot \left(\frac{dT_{air}}{dt}\right)_{max} \cdot V_{sphere^{1/3}} \tag{8.6}$$

where: $\left(\frac{dP}{dt}\right)_{max}$ is the maximum rate of pressure rise, n_{air} is the number of moles of air inside the sphere (considered fixed for the sake of simplicity) and, R is the ideal gas constant.

A second interesting approach for predicting K_{st} for organic powders is provided by Scotton et al. 2020 [54]. The peculiarity introduced by this validated model compared to previous ones is the inclusion of the entire particle size distribution (PSD) of the powder in the calculation process, whereas until the time of its publication, the common approach was to use the mean D50 value to represent the size of all powder particles included in the analysis. The choice to explore this physical aspect of the particles stems from the fact that each size class of the powder contributes differently to the characterization of the powder explosivity, including the deflagration index. Therefore, the K_{st} value is strongly influenced by the PSD. This occurs according to the simplification that the smaller the mean characteristic particle size, the higher the K_{st} value.

The core principle of this model is, therefore, based on the hypothesis that the K_{st} value of an organic powder with an arbitrary PSD can be estimated using the weighted average of all individual diameters that make up the PSD, according to Equation (8.7):

$$K_{st}(PSD) = \sum_{j=1}^{N} K_{st}(D_j) \cdot \omega(D_j) \tag{8.7}$$

where: $K_{st}(PSD)$ is the deflagration index value for an organic powder with a PSD characterized by the weight of the j-th particle size class $\omega(D_j)$. By dividing Equation (8.7) by the analogous equation for estimating K_{st} for a powder with a known PSD

(K_{st} (PSD_{ref})), it is possible to eliminate any reference to the D50 parameter, and the adopted approach allows the chemical reaction kinetics of the process to be disregarded, since they are included in the experimental K_{st} (PSD_{ref}) value of the equation analogous to Equation (8.7). Knowing the PSD value of a reference powder (with measured K_{st}) and the PSD value of a new powder for which the K_{st} value is to be estimated, it is then straightforward to estimate $\omega(D_j)$ and $\left.\frac{\partial T_{air}\left(D_j\right)}{\partial t}\right|_{max}$, the latter, obtained by solving Equation (8.5), and to calculate the $K_{st}(PSD)$ value [54].

8.2.4 Research Prospective

The possible future of Process Safety has the potential to encompass a vast range of scientific and humanistic fields capable of integrating with each other, with the aim of improving the current state of safety systems and measures. It is, therefore, reasonable to assume that new combinations of disciplines will add to those already existing, where engineering, psychology, natural sciences, computer science, and communication interact with the common goal of enhancing the systems and tools for prevention and protection against industrial accidents. Within this vast array of possible connections, however, it is useful to identify some that are representative of the need to guide the evolution of industrial safety techniques and tools toward meeting the requirements posed by the new challenges that the market and technology present at an ever-increasing pace.

A recent study on research trends in the field of dust explosions suggests that potential investigations could be conducted on nanopowders [55], to which hybrid mixtures, scale-up, and venting systems can be added.

8.2.4.1 Nanopowders

The presence of nanopowders in the food and pharmaceutical industries has increased over the past decade, also thanks to the spread of nanotechnologies, and the growing interest in this specific category of powders is demonstrated by the ever-increasing number of publications, patents [56], and market analyses [57] concerning them, with a projected steady growth in both academic studies and industrial applications.

The food sector is mainly interested in applications related to the improvement of packaging through the use of metallic nanoparticles to enhance their protection and its antimicrobial effect [58]. Moreover encapsulation of vitamins, antioxidants, and flavors using nanoparticles can serves to increase their bioavailability or stability of the product [59].

The applications of nanopowders and nanoparticles in the pharmaceutical sector are, mainly, the following: applications in oral and inhalation spray formulations to increase the bioavailability of the active ingredient [60], development of controlled release systems in which polymeric, lipid-based, or metallic nanoparticles can be used as drug carriers [61].

The presence of nanopowders and nanoparticles in the food and pharmaceutical industries is, therefore, significant, and this also creates the need to analyze their contribution in terms of increased industrial safety risks, such as those related to the phenomenon of dust explosions.

Nanoparticles are defined as materials with dimensions between 1 and 100 nm, which exhibit unique physicochemical properties due to their size. They can be composed of different chemical substances and can assume various morphologies, such as spheres, cylinders, or tubes [62]. These characteristics make it clear that the small size of nanoparticles is a factor that results in a high surface-to-volume ratio for each particle, significantly increasing their reactivity [63], along with a tendency to remain suspended longer than larger particles [64].

It is useful to remember that particle size of a powder is a factor that greatly influences both the powder ignition sensitivity and the magnitude of the explosion that may develop, since the pyrolysis phenomenon is itself affected by the particle size. Nanoparticles, therefore, are generally characterized by a lower MIE compared to powders with sizes on the order of micrometers [65] and by a K_{st} that increases as the particle size decreases, as long as it remains within the micrometer range, while no substantial differences are observed when transitioning to the nanometer range [66] and these properties create the perfect mix for a potential severe dust explosion.

At present, it appears that the possibilities for advancing the knowledge of the explosivity parameters of nanopowders are limited by the capabilities of testing instruments, which, based on their design specifications, are primarily calibrated to test micropowders. Similarly, testing standards mostly refer to micropowders, even though their explosive performance differs from that observed for nanopowders. In this context, and based on potential further developments useful for more detailed understanding of the reactivity and explosivity of nanopowders, it could be beneficial to implement studies and research aimed at establishing a more specific Particle Size Distribution (PSD) analysis procedure, designing more suitable explosivity testing instruments and standards, developing tailored suppression systems, and creating ad hoc explosion risk analysis procedures, thereby optimizing safety both during storage and throughout the production process in the presence of nanopowders [64].

8.2.4.2 Hybrid Mixtures

The presence of hybrid mixtures of powders and flammable gases/vapors is relevant in both the food industry and, even more so, in the pharmaceutical industry [67], and the risk that these mixtures may cause industrial explosions is higher than that associated with either powder or gas/vapor alone [68].

The relevant European standard for the definition of hybrid mixtures is CEI EN 60079-10-2:2016, which, in Annex C, specifies that these are "combined mixtures of flammable gas or vapor with a combustible powder or combustible particles in which a gas concentration $\geq$ 25% of the LFL value or a powder concentration $\geq$ 25% of the MEC value determines the explosivity of the mixture." According to the definition reported in the NFPA 660 standard, on the other hand, hybrid mixtures are "explosive heterogeneous mixtures with suspended solid or liquid particles, in which the gas concentration is $\geq$ 10% of the LFL value and the concentration of suspended particles is $\geq$ 10% of the MEC value." Essentially, this means that the presence of a flammable gas/vapor influences the explosivity of the pure powder to such an extent that the mixture becomes flammable and explosive even if the concentrations of gas and powder are each below their respective LFL and MEC values.

The reference values of MEC and LFL cited for the NFPA 660 standard and for the CEI EN 60079-10-2:2016 mainly derive from three macro-research blocks, such as:

- Experimental evidence. Various research studies available in the literature demonstrate that the ignition of a mixture composed of gases/vapors and powders can lead to an explosive phenomenon even when each component is characterized by explosivity values (in this case LFL and MEC) lower than those associated with its pure state [69, 70, 71].
- Variability of experimental data. The results provided by explosivity tests conducted in the laboratory are always subject to a certain margin of error, which consequently leads to variability in the produced data. It is, therefore, not simple to standardize the limits that can be used to define safe and unsafe values within a narrow range. For this reason, committees such as IEC and NFPA may adopt different approaches rather than using a single universally agreed-upon formula to determine precise safety limits [68–72].
- Applicability. The various committees involved in the development of standards that define safety thresholds establish limits that can ensure safety while also being realistically applicable in different contexts, such as the design of safety devices and preventive analyses. The differing values reported in NFPA 660 and CEI EN 60079-10-2:2016, while considering what has been previously mentioned, may also be influenced by the different industrial and regulatory contexts in which these two standards are adopted (NFPA 660, CEI EN 60079-10-2:2016).

In summary, therefore, the values reported in the standards are the result of a multifactorial combination rather than the result of a universally shared physical law. In this regard, it is, therefore, reasonable to ask what the explosive potential and propensity might be for gas/vapor and powder mixtures that, by definition, are not considered hybrid mixtures. More specifically, the question could be: do gas/vapor and powder mixtures that, by definition, cannot be classified as hybrid mixtures because their LFL and MEC values are below those reported in the standards, still possess an explosivity different from that of the pure powder? This issue aims to draw attention to the risk posed by mixtures that, although they exist, cannot be defined as hybrid mixtures because they do not meet the legal requirements. According to the work of Khalili et al. (2012) [73], it was demonstrated that the presence of 1% hexane in starch reduces the MIE of the powder from 67 mJ to 4 mJ, and that the gradual reduction of vapor concentration in the hybrid mixture, relative to the LEL of hexane, still affects the MIE value of the starch. It would, therefore, be interesting to explore this gray area where the "non-hybrid mixtures" reside; that is, mixtures that cannot be defined as hybrid from a regulatory standpoint, but still consist of the simultaneous presence of gases/vapors and powders. The need arises from the fact that the explosivity and magnitude of the non-hybrid mixture are still heightened compared to the pure powder, and with a view to possible future developments, it could be useful to develop a specific standard for both hybrid and non-hybrid mixtures, thereby addressing a condition that is often overlooked but relevant in the context of dust explosions.

8.2.4.3 Scale-Up

The use of laboratory instruments is now a well-established practice for characterizing the explosivity of powders present in the process industry. The values obtained through physical testing are the most reliable compared to any other possible alternative, making this approach the preferred choice in terms of result quality [74]. Nonetheless, it remains necessary to consider that the results obtained need to be scaled from a contained and controlled environment, such as the laboratory, to a larger one, the industrial setting, where the reactive phenomena involved in dust explosions are more complex. In this context, it is, therefore, necessary to critically analyze the scale-up process to maintain as much as possible the level of reliability provided by laboratory test results. In other words: *"if it were at all possible to produce a "perfect" experimental dust cloud, would this help us predict the course of real industrial dust explosions?"* [75]. It is, therefore, relevant that, for example, the estimation of a K_{st} value in the laboratory is representative of the risk associated with the specific powder in that context, without being able to adequately describe the dynamics of a dust explosion on an industrial scale where turbulence is different [76].

Even assuming that the powder used for laboratory tests faithfully represents the conditions of use in the real process, thus excluding possible alterations, there are still a series of aspects that should be considered in the scale-up phase of the data obtained in the laboratory. Among these, there are:

- Effect of dispersion. The homogeneous and repeatable dispersion required during laboratory testing does not represent the scenario that may occur in the event of a real dust explosion. In industrial plants, for example, areas of higher or lower dust concentration can form, influenced by the layout of the spaces in which the cloud can move, with consequent effects on the temporary availability of concentrations above the MEC value. Similarly, areas where the turbulence of the cloud differs from other zones and changes over time can affect the local magnitude of the explosion [75].
- Effect of geometry. The results obtained, for example, with the 20L sphere test do not account for the complex geometry present in actual industrial production plants. The presence of obstacles, venting devices, connections to secondary volumes, and the irregular geometry of certain structures strongly influence the development of the pressure wave on a real scale compared to a laboratory test environment. An assessment of the impact of these elements on the actual behavior of a dust explosion could be achieved through the use of CFD models [77].

- Effect of the ignition source. The ignition sources used in laboratory tests are controlled and deliberate discharges [78, 79], whereas real ignition sources are accidental, of various types (mechanical, thermal, electrical, chemical, or a possible combination thereof) and fall within a wide range of values capable of igniting the dust cloud [80].
- Secondary effects. In laboratory tests, it is not possible to include the cascading effect that a primary explosion exerts on a secondary explosion. As a result, the K_{st} parameters obtained cannot be representative of a real accidental scenario, which will, therefore, be more intense than what is measured during the tests [81].

Added to all this is the presence of intrinsic and inevitable errors associated with any type of measurement, which introduce a certain degree of uncertainty. It is, therefore, not advisable to use a single value as representative of an entire scenario without at least applying a confidence interval, especially when scaling to environments different from the test setting [82]. In practice, laboratory tests provide information on the intrinsic explosive properties of the powder under specific test conditions, but they cannot account for field variables. The development of a specific procedure to adjust laboratory values, perhaps through digital simulations or intermediate-scale steps, could be the subject of future studies aimed at optimizing Process Safety as applied to dust explosions.

8.2.4.4 Venting System

As a complement to prevention systems, which remain the primary line of defense against dust explosions, the use of protection systems, such as venting devices, completes the set of tools available to companies to counteract the phenomenon of dust explosions [83].

Dust collectors, silos, mills, grinders, process ovens, and pneumatic transport systems are just some of the elements in the food and pharmaceutical process industries that are protected by venting devices against overpressures generated by dust explosions [84–86]. This highlights the importance of both sectors of implementing the reliability of these protection systems through widespread optimization, from the design phase to installation and maintenance. However, it is crucial to emphasize that installing such devices is not recommended when dealing with powders or other potentially toxic substances that could be released into the atmosphere after the venting system is activated [87].

The sizing procedure for venting devices, as regulated by reference standards, is primarily based on the properties of the powder (K_{st} and P_{max}) and the characteristics of the enclosure on which the panel is to be installed [88–90]. In this regard, and in support of future implementations, it is worth noting that some companies may not have access to updated or sufficiently detailed analyses of the explosivity parameters of the powders they handle, which can lead to suboptimal sizing of protection systems for the specific case. If the data were updated and detailed, the use of scale-up models and AI systems could support the optimization of the device design.

In general, reference standards also require periodic inspection of the condition and compliance of overpressure venting devices, to which additional checks can be added to ensure their proper functioning and protective effectiveness [88–90]. Since regular maintenance and inspection of equipment in contact with combustible powders is one of the key elements in preventing the initiation of a dust explosion [91] and it is essential for ensuring the correct operation of the device [92], it could be useful, in this context, to implement the use of intelligent and interconnected devices (IIoT and ML) on venting systems (smart vent) for routine maintenance procedures, thereby optimizing the inspection process and adapting any interventions to the actual needs of the system.

REFERENCES

1. Cloney C., Barret J., 2025. NFPA 660 standard for combustible dust and particulate solids proposed. dustsafetyscience.com, March 2025. https://dustsafetyscience.com/nfpa-660-standard-for-combustible-dust/, last accessed 01/15/2026.
2. Maniar M.S., Kumar A., Mentzer R.A., 2020. Global process safety incidents in the pharmaceutical industry. *Journal of Loss Prevention in The Process Industries*, 68, 104279.
3. Boullenois C., Black M., Rosen D.H., 2025. Was made in China 2025 – Successful? Prepared for the US Chamber of Commerce 2025, Rhodium Group LLC.
4. KumarS.,MeenaS.,KaliaS.,2024.10yearsofmakeinIndiatransformingIndiaintoaglobal manufacturingpowerhouse. Ministry of Commerce & Industry of India, September 2024. https://static.pib.gov.in/WriteReadData/specificdocs/documents/2024/sep/doc2024925401801.pdf, last accessed 01/15/2026.
5. Indelicato M., 2025. L'india vuole diventare il maggior produttore di farmaci del mondo. *FEDAIISF*, July 2025. https://www.fedaiisf.it/lindia-vuol-diventare-il-maggior-produttore-di-farmaci-del-mondo/, last accessed 01/15/2026.
6. Brandt L., Rawski T.G., Sutton J., 2007. *China's Industrial Development in China's Great Economic Transformation*. Cambridge University Press, New York.
7. Liu T., Zhong M., Xing J., 2005. Industrial accidents: Challenges for China's economic and social development. *Safety Science*, 43, 503–522.
8. Tomasi G., 2011. La Sicurezza sul Lavoro in Cina con Case Study sulla Sicurezza nelle Miniere di Carbone. Master's Degree Thesis in Lingue e Istituzioni Economiche e Giuridiche dell'Asia e dell'Africa Mediterranea. Ca' Foscari University.
9. Meng Y., Hu Y., Li W., Liu Y., Zhao D., Willey R.J., 2023. The growth of Process Safety practice in China. *Process Safety Progress*, 42(1), 4–11.
10. Pang L., Zhang M., Yang K., Sun S., 2023. Scenario derivation and consequence evaluation of dust explosion accident based on dynamic Bayesian network. *Journal of Loss Prevention in the Process Industries*, 83(2), 105055.
11. Evans D., 2024. China's crossroads: Challenges & opportunities for the world's factory. Forbes. https://www.forbes.com/sites/daveevans/2024/11/26/chinas-crossroads-challenges--opportunities-for-the-worlds-factory/, last accessed 01/15/2026.
12. Teekah E., Balakrishnan P., 2025. Economy of India. Britannica Money, March 2025. https://www.britannica.com/money/economy-of-India, last accessed 01/15/2026.
13. Pallavi D., Pallabi D., Pallabi B., Debashis B., Pallavi S., 2024. Scenario of food processing sector in India. *Current Trends in Agriculture & Allied Sciences* (Volume 2), 638–645.

14. https://ispe.org/pharmaceutical-engineering/march-april-2025/indian-pharmaceutical-industry-creating-global-impact, last accessed 01/15/2026.
15. Vijayaraghavan G., 2010. Dust explosion in process industry: A case of greater awareness in India. *Journal of the Institution of Public Health Engineers*, 10, 37–43.
16. Italian Trade Agency ITA, 2025. India's food processing sector projected to reach $535 bn by fy26 end. ICE agenzia per la promozione all'estero e l'internazionalizzazione delle imprese italiane, September 2025. https://www.ice.it/it/news/notizie-dal-mondo/291396, last accessed 01/15/2026.
17. Shekhar C.I., 2024. Indian pharmaceutical industry: Past, present, and future outlook (2024–2030). DOI: 10.13140/RG.2.2.32592.01284.
18. Cloney C., 2022. Explosion at sugar factory in India kills two people and injures six. dustsafetyscience.com, August 2022. https://dustsafetyscience.com/sugar-dust-explosion-kakinada-india/, last accessed 01/15/2026.
19. Cloney C., 2018. Morning explosion at cake factory in India leaves 10 workers injured. dustsafetyscience.com, December 2018. https://dustsafetyscience.com/flour-dust-explosion-kolkata-india/, last accessed 01/15/2026.
20. Swathi V., 2025. Dust hazard analysis not mentioned in factories checklist. The Hindu, July 2025. https://www.thehindu.com/news/national/telangana/dust-hazard-analysis-not-mandatory-in-the-factories-checklist/article69784476.ece, last accessed 01/15/2026.
21. Rupasinghe W., 2025. Indian pharmaceutical factory explosion leaves 42 dead. Worls Socialist Web Site WSWS.org, July 2025. https://www.wsws.org/en/articles/2025/07/08/xygm-j08.html, last accessed 01/15/2026.
22. Bortolami V.S., 2022. L'industria farmaceutica indiana. Master's Degree Thesis in Istituzioni Economiche e Giuridiche dell'Asia e dell'Africa Mediterranea. Università Ca' Foscari, Venezia.
23. Redaction, 2024. Industria 5.0: come evolveranno le aziende. Team System Magazine, April 2025. https://www.teamsystem.com/magazine/industry-40/industria-5-0-evoluzione-aziende/, last accessed 01/15/2026.
24. Arunthavanathan R., Sajid Z., Amin M.T., Tian Y., Khan F., Pistikopoulos E., 2024. Process Safety 4.0: Artificial intelligence or intelligence augmentation for safer process operation? *AIChE Journal*, 70(7), e18475.
25. Park J., Kang D., 2024. Artificial intelligence and smart technologies in safety management: A comprehensive analysis across multiple industries. *Applied Science*, 14, 11934.
26. Pochmara J., Świetlicka A., 2023. Cybersecurity of industrial systems — A 2023 report. *Electronics*, 13, 1191.
27. Pasman J.H., Behie W.S., 2024. The evolution to Industry 5.0 / Safety 5.0, the developments in society, and implications for industry management. *Journal of Safety and Sustainability*, 1, 202–211.
28. Cloney C., Barrett J., 2024. Biomass dust explosion risks: Prevention and safety strategies. dustsafetyscience.com, September 2024. https://dustsafetyscience.com/biomass-dust-explosion/, last accessed 01/15/2026.
29. Miorandi D., Sicari S., De Pellegrini F., Chlamtac I., 2012. Internet of things: Vision, applications and research challenges. *Ad Hoc Networks*, 10, 1497–1516.
30. Gahan D., 2023. Internet of Things and its implications for the process industries. The Chemical Engineer, September 2023. https://www.thechemicalengineer.com/features/internet-of-things-and-its-implications-for-the-process-industries/, last accessed 01/15/2026.
31. Misra S., Roy C., Sauter T., Mukherjee A., Maiti J., 2022. Industrial internet of things for safety management applications: A survey. *IEEE Access*, 10, 83415–83439.
32. Bragatto P., Faramondi L., Failla F., Gnoni M.G., 2018. Potential and limits of IOT for hazardous job in process industries. *Chemical Engineering Transactions*, 67, 865–870.

33. Cloney C., Barrett J., 2025. How to prevent a combustible dust explosion: The dust safety science ultimate guide (2025). dustsafetyscience.com, January 2025. https://dustsafetyscience.com/how-to-prevent-a-combustible-dust-explosion/#role-of-automation-and-IOT, last accessed 01/15/2026.
34. Chandru P., Vadivel R., 2024. IoT-based intelligent drainage and dust identification system for enhanced hygiene and efficiency. *World Journal of Advanced Research and Reviews*, 21(3), 1789–1797.
35. Zakri M.B., Zamzami O., Babour A., 2024. Automatic dust reduction system: An IoT intervention for air quality. *International Journal of Advanced Computer Science and Applications*, 15(2), 173–180.
36. Xing L., Hongchao Y., Huigang S., Wenqi D., Ting Z., 2025. Internet of Things-driven intelligent ventilation system for buildings in humanized designs. *Applied Mathematics and Nonlinear Sciences*, 10, 1.
37. Moleda M., Małysiak-Mrozek B., Ding W., Sunderam V., Mrozek D., 2023. From corrective to predictive maintenance—A review of maintenance approaches for the power industry. *Sensors*, 23, 5970.
38. Daoutidis P., Lee H.J., Rangarajan S., Chiang L., Gopaluni B., Schweidtmann A.M., Harjunkoski I., Mercangoz M., Mesbah A., Boukouvala F., Lima F.V., Chanona A.R., Georgakis C., 2023. Machine learning in process systems engineering: Challenges and opportunities. *Computers and Chemical Engineering*, 181, 108523.
39. Walke S., Bakal J.W., 2024. Applicability of machine learning in process industry: A review. International Conference on Computing, Power, and Communication Technologies (IC2PCT).
40. Cloney C., Barrett J., 2025. Worst case analysis engineering and combustible dust explosions. dustsafetyscience.com, January 2025. https://dustsafetyscience.com/worst-case-analysis-engineering/, last accessed 01/15/2026.
41. Sattari F., Kurian D., Lefsrud L., Macciotta R., 2021. Using artificial intelligence and machine learning techniques to analyze incident reports. HAZARDS 31, SYMPOSIUM SERIES, 168.
42. Redaction, 2024. Sensor-based predictive maintenance of equipment installed in explosive atmospheres. IoTforall.com. https://www.iotforall.com/sensor-based-predictive-maintenance-of-equipment-installed-in-explosive-atmospheres-ex-certified?utm, last accessed 01/15/2026.
43. Shen X., Zhang Z., Ma Y., Zou X., Zhou F., Wang S., Bao Q., 2023. Prediction of minimum ignition energy for combustible dust using graph neural networks and SMILES data augmentation. *Powder Technology*, 429, 118876.
44. Yu D., He Z., 2022. Digital twin-driven intelligence disaster prevention and mitigation for infrastructure: Advances, challenges, and opportunities. *Natural Hazards*, 112, 1–36.
45. Yu X., Li X., 2023. Sound identification method for gas and coal dust explosions based on MLP. *Entropy (Basel)*, 25(8), 1184.
46. Aslam M.M., Tufail A., Gul H., Irshad M.N., Namoun A., 2025. Artificial intelligence for secure and sustainable industrial control systems – A survey of challenges and solutions. *Artificial Intelligence Review*, 58, 349.
47. Velten K., 2009. *Mathematical Modeling and Simulation – Introduction for Scientists and Engineers*. WILEY-VCH Verlag GmbH & Co. KGaA, Weinheim.
48. Haines C., Crouch R., 2007. Mathematical modelling and applications: Ability and competence frameworks. In: Blum W., Galbraith P.L., Henn HW., Niss M., *Modelling and Applications in Mathematics Education*. New ICMI Study Series, vol 10. Springer, Boston, MA.

49. Verschaffel L., Greer B., De Corte E., 2002. Everyday knowledge and mathematical modeling of school word problems. In: *Symbolizing, Modeling and Tool Use in Mathematics Education*. Springer, Berlin, 257–276.
50. Copelli S., Scotton M.S., Barozzi M., Derudi M., Rota R., 2021. A practical tool for predicting the minimum ignition energy of organic dusts. *Industrial & Engineering Chemistry Research*, 60(29), 10807–10813.
51. Copelli S., Barozzi M., Scotton M.S., Fumagalli A., Derudi M., Rota R., 2019. A predictive model for the estimation of the deflagration index of organic dusts. *Process Safety and Environmental Protection*, 126, 329–338.
52. Copelli S., Barozzi M., Ballinari D., Lugaresi O., Moneta R., 2025. Application of critical energy density concept for minimum ignition energy determination. *Journal of Electrostatics*, 136, 104131.
53. Mencik M., Kuracina R., Szabova Z., Balog K., 2016. Determination of fire and explosion characteristics of dust. Safety Engineering Series, 36–42.
54. Scotton M.S., Barozzi M., Derudi M., Rota R., Copelli S., 2020. Kinetic free mathematical model for the prediction of Kst values for organic dusts with arbitrary particle size distribution. *Journal of Loss Prevention in the Process Industries*, 67, 104218.
55. Guan W., Jin M., Dong C., Gong H., 2023. Analysis on research trends with dust explosions by bibliometric approach. *Journal of Loss Prevention Industries*, 81, 104958.
56. WHO and FAO, 2010. FAO/WHO Expert meeting on the application of nanotechnologies in the food and agriculture sectors: Potential food safety implications meeting report. Food and Agriculture Organization of the United Nations and World Health Org, ISBN 978-92-4-156393-2 (WHO), ISBN 978-92-5-106506-8 (FAO). https://www.fao.org/4/i1434e/i1434e00.pdf, last accessed 01/15/2026.
57. Redaction, 2025. Nanomaterials market (2026 - 2033). grandviewresearch.com. https://www.grandviewresearch.com/industry-analysis/nanotechnology-and-nanomaterials-market, last accessed 01/15/2026.
58. Couto C., Almeida A., 2022. Metallic nanoparticles in the food sector: A mini-review. *Foods*, 11(3), 402.
59. Mallia J.O., Galea R., Nag R., Cummins E., Gatt R., Valdramidis V., 2022. Nanoparticle food applications and their toxicity: Current trends and needs in risk assessment strategies. *Journal of Food Protection*, 85(2), 355–372.
60. Ozturk A.A., Arpagaus C., 2021. Nano spray-dried drugs for oral administration: A review. *ASSAY and Drug Development Technologies*, 19, 7.
61. Ewii U.E., Attama A.A., Olorunsola O.E., Onugwu A.L., Nwakpa F.U., Anyiam C., Chijioke C., Ogbulie T., 2025. Nanoparticles for drug delivery: Insight into in vitro and in vivo drug release from nanomedicines. *Nano TransMed*, 4, 100083.
62. Joudeh N., Linke D., 2022. Nanoparticle classification, physicochemical properties, characterization, and applications: A comprehensive review for biologists. *Journal of Nanobiotechnology*, 20, 262.
63. EFSA, 2025. Nanotechnology. European Food Safety Authority. https://www.efsa.europa.eu/en/topics/topic/nanotechnology, last accessed 01/15/2026.
64. Santandrea A., Vignes A., Krietsch A., Brunello D., Perrin L., Laurent A., Dufaud O., 2020. Evaluating the explosion severity of nanopowders: International standards versus reality. *Process Safety and Environmental Protection*, 138, 279–291.
65. Dobashi R., 2009. Risk of dust explosions of combustible nanomaterials. *Journal of Physics: Conference Series*, 170, 012029.
66. Rabajczyc A., Wyszomirska M., Zboina J., 2024. Selected aspects of the use of nanotechnology – Solutions and challenges in the field of safety and in the regulatory area. *Scientific Reports of Fire University ZN SGSP*, 1(89), 63–84. DOI: 10.5604/01.3001.0054.4248. https://zeszytynaukowe-sgsp.pl/article/544248/en, last accessed 01/15/2026.

67. Amyotte P., Khan F.I., 2019. *Methods in Chemical Process Safety*, 3rd volume. Academic Press Publications.
68. Wu D., Zhao P., Spitzer S., Krietsch A., Amyotte P., Krause U., 2023. A review on hybrid mixture explosions: Safety parameters, explosion regimes and criteria, flame characteristics. *Journal of Loss Prevention in the Process Industries*, 82, 104969.
69. Jiang J., Liu Y., Mannan M.S., 2014. A correlation of the lower flammability limit for hybrid mixtures. *Journal of Loss Prevention in the Process Industries*, 32, 120–126.
70. Sanchirico R., Di Benedetto A., Garcia-Agreda A., Russo P., 2011. Study of the severity of hybrid mixture explosions and comparison to pure dust–air and vapour–air explosions. *Journal of Loss Prevention in the Process Industries*, 24(5), 648–655.
71. Addai E.K., Addo A., Abbas Z., Krause U., 2017. Investigation of the minimum ignition temperature and lower explosion limit of multi-components hybrid mixtures in the Godbert-Greenwald furnace. *Process Safety and Environmental Protection*, 111, 785–794.
72. Jiang J., 2015. Study of dust – Gas hybrid mixture explosions. Dissertation of PhD Thesis in Chemical Engineering. Texas A&M University.
73. Khalili I., Dufaud O., Poupeau M., Cervo-Rodriguez, Perrin L., 2012. Ignition sensitivity of gas–vapor/dust hybrid mixtures. *Powder Technology*, 217, 199–206.
74. Barozzi M., Ballinari D., Scotton M.S., Copelli S., 2025. Connecting standards and theoretical models for minimum ignition energy estimation: A procedure to optimize the number of experimental tests. *Journal of Electrostatics*, 138, 104196.
75. Eckhoff R.K., 2015. Scaling of dust explosion violence from laboratory scale to full industrial scale e A challenging case history from the past. *Journal of Loss Prevention in the Process Industries*, 36, 271–280.
76. Regis C., Bauwens L., Boeck L.R., Dorofeev S.B., 2024. Characterizing the reactivity of large-scale dust explosions with a dimensionless two-parameter model. *Proceedings of the Combustion Institute*, 40(1–4), 105280.
77. Skjold T., Arntzen B.J., Hansen O.R., Storvik I.E., Eckhoff R.K., 2006. Simulation of dust explosions in complex geometries with experimental input from standardized tests. *Journal of Loss Prevention in the Process Industries*, 19(2–3), 210–217.
78. UNI EN 14034--1:2011 "Determinazione delle caratteristiche di esplosione di nubi di polvere."
79. EN ISO/IEC 80079-20--2:2016 Explosive atmospheres – Part 20-2: Material characteristics – Combustible dusts test methods.
80. Cloney C., Barrett J., 2025. What are ignition sources in industrial safety and dust explosions. dustsafetyscience.com, March 2025. https://dustsafetyscience.com/ignition-sources/#A2, last accessed 01/15/2026.
81. Eckhoff R.K., Li G., 2021. Industrial dust explosions. A brief review. *Applied Science*, 11, 1669.
82. Taylor J.R., 1986. *Introduzione all'analisi degli errori: lo studio delle incertezze nelle misure fisiche*. Zanichelli, Bologna.
83. Snoeys J., Going J.E., Taveau J.R., 2012. Advances in dust explosion protection techniques: Flameless venting. *Procedia Engineering*, 45, 403–413.
84. OSHA, 2020. Technical manual – Section IV, chapter 6, combustible dusts. Directive Number TED-01-00-015.
85. Holbrow P., Andrews S., Lunn G.A., 1996. Dust explosions in interconnected vented vessels. *Journal of Loss Prevention in the Process Industries*, 9(1), 91–103.
86. ISSA International Social Security Association, 2013. "Collection of examples: Dust explosion protection for machines and equipment – Part 1." Published by ISSA.

87. Mannan M.S., 2012. *Lees' Loss Prevention in the Process Industries: Hazard Identification, Assessment and Control: Fourth Edition.*, Butterworth-Heinemann, Oxford (UK).
88. NFPA 68, 2023. "Standard on explosion protection by deflagration venting."
89. EN 14491-2012 Dust explosion venting protective systems.
90. EN 14797-2006 Explosion venting devices.
91. Donaldson Company, 2025. Combustible dust brochure. Dust Collectors and Combustible Dust Management. https://www.donaldson.com/content/dam/donaldson/dust-fume-mist/literature/emea/equipment/combustible-dust/f118178/euk/Combustible-Dust-Brochure.pdf, last accessed 01/15/2026.
92. Cloney C., Barrett J., 2024. A complete guide to an explosion vented equipment system. dustsafetyscience.com, August 2024. https://dustsafetyscience.com/explosion-vented-equipment-system/, last accessed 01/15/2026.

Index

A

B

C

D

V

W

X

Z